CLIMATE CHANGE AND ENERGY OPTIONS FOR A SUSTAINABLE FUTURE

CLIMATE CHANGE AND ENERGY OPTIONS FOR A SUSTAINABLE FUTURE

Dinesh Kumar Srivastava
V S Ramamurthy
National Institute of Advanced Studies, India

NEW JERSEY • LONDON • SINGAPORE • BEIJING • SHANGHAI • HONG KONG • TAIPEI • CHENNAI • TOKYO

Published by

World Scientific Publishing Co. Pte. Ltd.
5 Toh Tuck Link, Singapore 596224
USA office: 27 Warren Street, Suite 401-402, Hackensack, NJ 07601
UK office: 57 Shelton Street, Covent Garden, London WC2H 9HE

British Library Cataloguing-in-Publication Data
A catalogue record for this book is available from the British Library.

CLIMATE CHANGE AND ENERGY OPTIONS FOR A SUSTAINABLE FUTURE

ISBN 978-981-123-347-0 (hardcover)
ISBN 978-981-123-348-7 (ebook for institutions)
ISBN 978-981-123-349-4 (ebook for individuals)

For any available supplementary material, please visit
https://www.worldscientific.com/worldscibooks/10.1142/12187#t=suppl

Dedication

न त्वहं कामये राज्यं न स्वर्गं नापुनर्भवम्,
कामये दुःखतप्तानां प्राणिनामार्तिनाशनम् ||

"I desire neither crown, nor heaven, nor even salvation; the only desire I have is that the sorrows of all those who are suffering comes to an end"

Reaching Out

The rise in global temperature and the climate change is the biggest long-term threat which mankind faces today. This is already leading to frequent heatwaves of longer durations, extreme weather events like very heavy rainfalls — accompanied with prolonged droughts, surge floods, rise in sea-level — which threatens drowning of coastal areas and island nations, frequent and fiercer cyclones, forest fires, crop failures, decimation of animals and marine life, increase in pests, and rise in infectious — especially water borne — diseases, and will lead to a large number of climate refugees, if it remains unchecked.

The Intergovernmental Panel on Climate Change has suggested that the climate change could be arrested by limiting the rise of global temperature to within 1.5–2.0 degree Celsius of the pre-Industrial Revolution values. This requires that the emission of greenhouse gases be reduced to 40% of the level of 2010, by 2030 and to zero by 2050.

The dual requirement of raising the per capita energy available to our people to a level of about 4000 kWh per year from the present value of about 1200 kWh per year for a decent life and arresting the emission of greenhouse gases from the use of fossil fuels has only one solution — to switch over to renewable and green sources of energy, immediately. National Institute of Advanced Studies has been at the forefront of policy studies for energy and other security issues, as well as in bringing the scientific solutions to the attention of policy makers and people of India.

The present book on "Climate Change and Energy Options for a Sustainable Development" by Dinesh K. Srivastava and V. S. Ramamurthy takes a holistic view of climate change — putting it in historical perspective and proceeding through the Industrial Revolution, which changed the World for ever. It discusses all the major sources of energy like coal, oil, gas, biomass, biofuel, hydroelectricity, solar power, windmills, hydrogen, and nuclear power along with their advantages, disadvantage, and availability. The discussion is very lucid and free from technical complexities, so that the book is accessible to young students as well as enlightened persons.

It is hoped that the book will play an important role in raising the scientific awareness of our people about the challenge of climate change and encourage them to welcome and support the corrective measures, whole heartedly and enthusiastically.

Prof. Shailesh Nayak
Director
National Institute of Advanced Studies
Bengaluru 560012

Foreword

Charles Dickens, started his novel, "A Tale of Two Cities", with the immortal words, "It was the best of times, it was the worst of times, it was the age of wisdom, it was the age of foolishness, it was the epoch of belief, it was the epoch of incredulity, it was the season of Light, it was the season of Darkness, it was the spring of hope, it was the winter of despair, we had everything before us, we had nothing before us, we were all going direct to Heaven, we were all going direct the other way, ...".

These words very aptly describe the predicament in which mankind finds itself following the all-round developments in sciences and technology, and the incredible improvement in the standards of living of common people, ushered in by Industrial Revolution but which has triggered climate change. Pericles, speaking in 430 BC at his famous funeral oration, remarked, "Because of the greatness of our city the fruits of the whole Earth flow in upon us; so that we enjoy the goods of other countries as freely as our own." This statement also acknowledged that globalization, of which we have enough proofs from Indus Valley, Sumerian, Akkadian, and Egyptian civilizations, was already well established. It was a consequence of man's desire to explore, conquer, exploit, learn, and trade — as well as to spread his world views, including religious beliefs. The globalization also led to a rapid dissemination of knowledge and accelerated the process of societal change. Today's world is rightly described as a global village.

By now we know that man started changing his surroundings by using fire to clear forests and then by agriculture. In fact, agriculture produced food surplus and left a part of population free to develop the arts, literature, music, writing, architecture, and sciences. However, several of these civilizations collapsed after dazzling the world with their brilliance. While some of these were lost to wars, many were lost to over-exploitation, which led to denuding of natural resources and even alteration of local climate or fertility of the soil. Perhaps, the changes were slow, and the people did not quite realize or ignored the shape of things to come.

Industrial Revolution resulted in an increase of "carrying capacity" of the Earth by several orders of magnitude. But it has also put its resources and the capacity of our planet to recover from the damage caused to its ecosystem by man and his activities under severe strain. It is generally accepted that the global warming and climate change being brought about by the accumulation of greenhouse gases in the atmosphere, indiscriminate mining, destruction of forests, pollution and over-exploitation of oceans and the lands, has brought us to a stage, from where we may never recover and the Earth may plunge into a hot house, if we continue in our ways.

And yet, the only way to pull all the mankind out of poverty, sickness, and lack of education is to provide them with more energy. This energy has mostly come from burning of fossil fuels so far, which is also responsible for the global warming.

The signs of global warming, as the authors of this book discuss, are all around us; "... the summers are getting hotter everywhere, heavy rains and prolonged droughts are recurring more frequently, forest fires are getting more severe, storms (including snow storms) and cyclones are more frequent and fierce, the polar ice caps are melting — raising the spectre of millions of climate refugees, the glaciers are dying — raising the spectre of wars for water, crops are failing, fish are dying, coral reefs are whitening." It is even more worrying that the poorer countries, which are least equipped to face these calamities and which contributed least to the global warming, are the worst hit.

What are the options before the mankind today, to face this challenge which may essentially reverse its technological gains and even threaten its very survival?

"Climate Change and Energy Options for a Sustainable Future" is an attempt to discuss these issues. It starts by giving historical perspectives of the collapse of some representative civilizations and destruction of our delicate biodiversity — which has evolved over billions of years, and which has recovered from natural calamities, including ice-ages and asteroid strikes. Then it proceeds to discuss energy options available to and availed by mankind; coal, oil, and natural gas, biomass and biofuel, wind, geothermal, tidal, hydroelectricity, solar photovoltaics, concentrated solar thermal, and nuclear power, along with their feasibility, availability, and their advantages and disadvantages. Hydrogen is discussed as an emerging option for carrier of clean energy. The authors have also discussed the culture of greed, fast fashion, and over-indulgence, which is resulting in serious over-exploitation. I find that the authors have kept the discussions detailed, but without burdening the reader with technical details. This is very important for the cultivation of an educated, well informed population, which through their choices and actions can bring about a balance between development and sustainability.

It is quite clear that to maintain a healthy planet, every possible effort is required of the global community to limit the increase in global temperature to less than 2 degrees Celsius, as recommended by Intergovernmental Panel for Climate Change. This will require that global carbon dioxide emissions must reduce by 40% of 2010 by 2030 and to net zero by 2050!

This requires rapid deployment of renewable energy sources, possibly with a mix of the much cleaner fossil fuel technologies where a complete abandoning of the latter is not possible in the near term. The authors suggest that nuclear power can be used to provide the base load, to account for the intermittency of the renewables. Storage ideas like pumped up hydroelectricity, batteries, and hydrogen are discussed. It is also suggested to use a combination of energy sources, e.g., aqua-photovoltaics — putting solar cells on hydroelectric dams and other water bodies, or covering large areas with solar photovoltaics and windmills, if feasible, or producing hydrogen using excess energy from renewables, *etc.*

The authors invoke the spirit of international cooperation and international collaboration, with a common vision for one Earth, which as they correctly point out can unleash a huge positive energy required to effectively tackle this unprecedented challenge faced by the entire mankind.

It is for the first time that we will have to think and act as truly global citizens to together manage and balance the needs of the planet with a clear and present shared vision, direction, and actions. Together we can.

Prof. Ashutosh Sharma
Secretary
Department of Science and Technology
Government of India

Preface

It was an early morning, more than forty years ago, in one of the new neighbourhoods of Kolkata. The city was still called Calcutta in those days. Two young scientists were taking a leisurely walk along a street lined with a few houses and still fewer trees. One of them (D K Srivastava) was a new migrant to Calcutta, having been posted at the newly established Variable Energy Cyclotron Centre and the other (V S Ramamurthy) had arrived to perform an experiment at the Cyclotron. The air was cool, but one could not escape a sharp smell of smoke. The reason was not hard to find. Till early eighties, the defining moments of mornings and evenings in cities like Kolkata were the preparations of a coal oven for cooking. A metallic bucket, with steel rods inserted across to make a grill, with an opening below the grill but above the bottom and a coating of mud or cement on the insides, was used as the oven. The contraption was called "angeethee" (a brazier). Raw coal, for which there used to be shops in every locality, was loaded in the oven above dry pieces of wood. The arrangement of the coal and the wood pieces was an art and allowed air to get in and the smoke to get out. A fire lit at the bottom of the oven using paper or wood would first light the wood above the grill and then slowly the coal would start smouldering. The resulting dense smoke, acrid and choking, loaded with particulate matter, sulphur dioxide and of course plenty of carbon dioxide and many other pollutants would fill the entire surrounding. The oven used to take almost 30 minutes to ignite all the pieces of coal to a red-hot stage without giving out smoke, when it was ready to be used for cooking. The smell of the sulphur was the permanent resident of the localities. The large-scale availability of cooking gas and electric ovens have rescued us from this "ritualistic" torture by choking.

Now imagine a coal thermal power plant, having a capacity of about 1000 MWe. It will need about four million tonnes of coal every year. It would be continuously spewing out smoke, carbon dioxide (14.7 million tons of it every year!), sulphur dioxide, nitrous oxide, smoke, particulate matter, and many other pollutants including ash and Uranium. And now imagine a world needing about 20 TWe of power. Up to 80% of this is generated from coal and natural gas. The world emitted 37 billion tons of carbon dioxide in 2019.

The industrialisation has pulled billions out of a life of abject poverty, disease, and suffering and given them a life which was open only to the members of aristocracy and extremely rich merchants. Now people of the world live healthier, longer and a lot more comfortably, even though about one-third of the people of the world still need to be pulled out of poverty. Rapid continuing developments in science and technology have made it possible. The recent developments in transport and communication technologies have also turned the World into a Global Village. Globalisation, which began with trade, is also leading to rapid dissemination of knowledge across the World and cooperation among the scientists and engineers of the World. This has led to an ever-accelerating pace of discovery and development.

All these developments have needed energy — well beyond what man can provide using his muscles. Once man had access to the concentrated chemical energy accumulated in coal over millions of years ago, the Industrial Era started. With the beginning of the Industrial Era in mid nineteenth century, the emission of carbon dioxide and other greenhouse gases has been on the rise. We have experimental measurements which confirm that during the 800,000 years before 1800, the level of carbon dioxide never rose above 300 parts per million (ppm), even though the Earth itself underwent many cycles of cold and hot periods. It has touched a value of 410 ppm. And ever since, the temperature of our planet has continued to rise. It has already risen by almost 1.1 degree Celsius and continues to rise. Scientists are convinced that unless we reduce carbon dioxide emissions to near zero by 2050 and limit the rise of the temperature to less than 1.5–2.0 degree Celsius, the stability of biodiversity and ecosystem of the Earth and survival of human-civilization may be seriously jeopardised. Even if we completely and abruptly stop the emission of carbon dioxide due to fossil fuel burning and industries today, the Earth will continue to get hotter for decades before it starts the process of recovery as carbon dioxide stays in the atmosphere for hundreds of years.

The warning signs are all around us — the summers are getting hotter everywhere, heavy rains and prolonged droughts are recurring more frequently, forest fires are getting more severe, storms (including snow storms) and cyclones are more frequent and fierce, the polar ice caps are melting — raising the spectre of millions of climate refugees, the glaciers are dying — raising the spectre of wars for water, crops are failing, fish are dying, coral reefs are whitening. Even more worrying fact is that the poorer countries, which contributed least to the global warming and which are least equipped to face these calamities, are the worst hit.

All these have greatly worried scientists and thinkers, who have been trying to raise these issues for decades. The immediate requirement of drawing people out of poverty in third world countries and to provide continued conspicuous consumption and the life of extreme comfort in advanced nations have led the governments to lean on easy solutions, and thus their dependence on fossil fuels has continued. Unfortunately, we cannot continue like this. The Earth and humanity are hurtling down a path of self-destruction at a break-neck speed.

The time to act is now

We know that when the entire world rises as one, it unleashes an unprecedented power and capability. We have witnessed several instances of this. Science has several solutions to offer. We discuss these in this book. This book arose out of discussions we had among ourselves and with others about various developments in the field of climate change and possible solutions as well as the exciting developments in various options for renewable energy. It was felt that if the common people and especially young students are made aware of the severity of the looming disaster and options available to us to meet this challenge, we shall have their wholehearted support to our efforts.

We realize that climate change affects the future of our children most strongly. They are obviously quite concerned about it and have regularly resorted to agitations across the world, following the example of Greta Thunberg of Sweden to draw the attention of the governments. It is sad that in spite of overwhelming scientific evidence, there are some people — even in positions of authority, who continue to deny climate change and cause severe harm to the efforts of the rest of the world towards its mitigation. The United Nations General Assembly President, María Fernanda Espinosa Garcés (Ecuador), noted in 2019 that, “We are the last generation that can prevent irreparable damage to our planet.” And we just have about ten years left to effect this change.

A knowledgeable and understanding populace is the biggest asset the democratic world has. This book is addressed to this enlightened populace. If it plays even a small role in generating a concern, an understanding and a support for optimal solutions for the crisis the entire world faces today, among people and most importantly among young students, we shall feel that our efforts have not been in vain.

Several persons have made this possible. We have benefitted from the valuable comments of S. A. Bhardwaj, Rupa Chatterjee, Ashok Kumar Jain, Rudrodip Majumdar, Angelica Muller, Berndt Muller, Shailesh Nayak, Harini Santhanam, R. K. Sinha, R. Srikanth, Devesh Srivastava, Lalitha Sundaresan, and numerous other persons. We also benefitted from the talks which were delivered on this topic at various forums and from the feedbacks received.

Both of us have used our spouses, Mrs. Rekha Srivastava and Mrs. Raji Ramamurthy respectively, as sounding boards to test our hair-brained ideas, approaches and feelings and have benefitted immensely from their earthy comments. We thank them for tolerating our excitement and involvement in bringing this effort to the present shape while neglecting our part of duties in running the family and the household.

We have enjoyed our discussions, our formulations of questions, our search for answers and our mutually corrective efforts while bringing out this book. We have continuously strived to make it understandable to an enlightened general populace and inquisitive students. This has necessitated skipping many technical details for which the experts should excuse us.

To keep our arguments as basic as possible, we have, subjectively taken, illustrations from publicly available websites and given references to them, so that an interested reader may get more details from there. We thank the contributors to those sites in advance.

DKS would like to put on record the support provided to him in the form of a position as Homi Bhabha Chair Professor by the National Institute of Advanced Studies, Bengaluru, and the very many valuable discussions with R. Srikanth, Professor and Dean, Energy and Environment Group as well as all the members of the group for their deep insights. VSR is grateful for the support provided by National Institute of Advanced Studies as an Emeritus Professor. We are extremely grateful to World Scientific Publishing Company, Singapore for readily offering to publish this work and for their help in arranging it in the final form.

We are grateful to Prof Ashutosh Sharma, Secretary Department of Science and Technology, Government of India for a critical reading of the manuscript and making valuable suggestions for its improvement. He has also, very kindly provided a Foreword for this book for which we are additionally grateful.

We are also grateful to Dr Shailesh Nayak, Director, National Institute of Advanced Studies, Bengaluru, for his continued support during the preparation of this manuscript and his valuable comments on an early draft of it. We are also grateful to him for his message "Reaching out".

Dinesh K Srivastava and V S Ramamurthy
Bengaluru, August 2020

Contents

Reaching Out vii
Foreword ix
Preface xiii
List of Tables xxi
List of Figures xxiii
About the Authors xxxi

1. Prologue 1
2. The World is Fragile 13
3. A Little More About Mother Earth 31
4. The Incredible Transformation of Hunter–Gatherers into Civilization Builders 51
5. Global Warming is for Real 73
6. Global Warming and Greenhouse Gases 85
7. Energy and Human Development Index 99
8. How Much Energy Do We Need? 105
9. Energy Resources 109
   9.1 Coal, Oil, and Natural Gas 110
   9.2 Biomass and Biofuel 125
   9.3 Wind Energy 137
   9.4 Geothermal Energy 140
   9.5 Tidal Energy 147
   9.6 Hydroelectric Energy 152
10. Solar Energy 161
   10.1 Solar Thermal 164

10.2 Solar Photovoltaics 180
11. Nuclear Power 205
12. Nuclear Fusion 243
13. Accelerator Driven Subcritical Systems 253
14. Nuclear Safety and Nuclear Waste 257
15. Healthcare and Other Applications of Nuclear Radiations 277
16. Hydrogen 289
17. Summary and Outlook 303
18. Epilogue 309

Bibliography 315
Index 327

List of Tables

Table 4.1. Power of Animals and Man 52

Table 6.1. Global Warming Potential of some Green House Gases 92

Table 8.1. Energy Demand Based on World Energy Statistics and Projections 106

Table 9.1. Pollutants Emitted by a 1300 MW Thermal Power Plant using Coal 135

Table 9.2. Top Five Hydroelectric Power Stations in the World 157

Table 9.3. Main Hydroelectric Power Stations of India 158

Table 10.1. Added and Total Solar Power (MW) in India 195

Table 10.2. Area Needed for Setting up Power Plants 200

Table 10.3. Total Area Needed for Renewable Energy Power Plants 201

Table 11.1. Fission Cross-Section and Prompt and Delayed Neutrons for Thermal and Fast Neutrons 218

Table 11.2. Thorium Reserves in the World 228

Table 11.3. Nuclear Reactors Operating in India 236

Table 11.4. Reactors under Construction in India 237

Table 14.1. Dose Limitations 263

Table 14.2. Radiation Dose from Natural Sources 263

Table 14.3. Radiation Dose from Various Activities 264

Table 16.1. Heat Value of Various Fuels 290

List of Figures

Fig. 1.1. The Sun Rising over a Desert ... 3
Fig. 1.2. A Rainbow Eucalyptus Tree in Australia ... 4
Fig. 1.3. An African Elephant ... 5
Fig. 1.4. A Royal Bengal Tiger ... 5
Fig. 1.5. An Indian Bison, Jaldapara ... 6
Fig. 1.6. A Bird of Paradise from Papua, New Guinea ... 6
Fig. 1.7. A Ulysses Butterfly ... 7
Fig. 1.8. A Humpback Whale Breaches Far in the Air above the Pacific Ocean ... 7
Fig. 1.9. A Dahlia ... 8
Fig. 1.10. Ra, the Sun god of ancient Egyptians ... 11
Fig. 1.11. The Inheritors of the Earth ... 12
Fig. 2.1. Congress Grass in India ... 16
Fig. 2.2. Angreji Babool in India ... 17
Fig. 2.3. Water Hyacinth ... 18
Fig. 2.4. The Erosion of a Gully in South Australia Caused by Overgrazing by Rabbits ... 19
Fig. 2.5. Nile Perch Introduced in Lake Victoria ... 20
Fig. 2.6. Myna in Sydney ... 21
Fig. 2.7. A Common Sparrow ... 21
Fig. 2.8. Babur Hunting Rhinoceros Near Peshawar ... 23
Fig. 2.9. Maharajah Ramanuj Pratap Singh Deo with the Last Three Cheetahs, 1948 ... 23
Fig. 2.10. Shooting of Passenger Pigeons ... 25
Fig. 2.11. Bees as Pollinators ... 26
Fig. 2.12. Hand Pollination of Apple Flowers in China ... 27
Fig. 2.13. A Handful of Soil May Contain Several Billion Bacteria ... 28
Fig. 2.14. Plastic Pollution in Yangtze River near Shanghai ... 29
Fig. 3.1. Earth Seen from Space ... 31
Fig. 3.2. Basic Building Blocks According to Ancient Greeks ... 32
Fig. 3.3. A Rabbit and A Rock ... 33
Fig. 3.4. Atomos of the Fur of a Lion and a Rock According to Democritus ... 33
Fig. 3.5. Dmitri Mendeleev's Periodic Table ... 34
Fig. 3.6. Dmitri Mendeleev and his Periodic Table ... 35
Fig. 3.7. Rutherford Atom ... 36
Fig. 3.8. The Basic Building Blocks of Matter ... 37
Fig. 3.9. History of the Early Universe ... 38
Fig. 3.10. Evolution of Life on Earth ... 39
Fig. 3.11. Life May Have Evolved Near Thermal Vents in Deep Oceans ... 41
Fig. 3.12. How is Life Sustained on Earth? ... 42

Fig. 3.13. Cross-section of the Earth ... 43
Fig. 3.14. Greenhouse Gases Heating the Earth ... 44
Fig. 3.15. Earth's Magnetic Field Protects it from Solar Flares ... 45
Fig. 3.16. Birth of the Moon ... 47
Fig. 3.17. Gondwana Land ... 48
Fig. 3.18. Fossils of Glossopteris Flora ... 48
Fig. 3.19. India arrives from Antarctica! ... 49
Fig. 4.1. Horsepower ... 52
Fig. 4.2. Oxen put to Work in India ... 53
Fig. 4.3. Travois used by Native Americans ... 53
Fig. 4.4. A Slave Market in Yemen ... 55
Fig. 4.5. Slaves from Africa being Taken to Americas ... 55
Fig. 4.6. Indentured Indian Labourers in Trinidad ... 56
Fig. 4.7. Pompei and Vesuvius ... 58
Fig. 4.8. Banda Aceh, Indonesia after a 9.0 Magnitude Earthquake (2004) ... 59
Fig. 4.9. The Nunnery Quadrangle in Uxmal ... 61
Fig. 4.10. Moai Statues in Easter Islands ... 62
Fig. 4.11. Sumerian Civilization ... 63
Fig. 4.12. The Great Bath of Mohenjo-Daro ... 65
Fig. 4.13. One of the early Steam Engines called John Bull (c. 1893) ... 67
Fig. 4.14. Annual Energy Consumption per Head (in Megajoules) in England and Wales ... 68
Fig. 4.15. World Primary Energy Consumption, 1950–2050 ... 70
Fig. 5.1. The Melting of Ice at Antarctica ... 74
Fig. 5.2. The Melting of Ice in the Arctic ... 74
Fig. 5.3. White Chuck Glacier in 1973 ... 75
Fig. 5.4. White Chuck Glacier in 2006 ... 75
Fig. 5.5. A Glacier in Tibetan Himalaya: 1921 (upper panel) and 2008 (lower panel) ... 76
Fig. 5.6. The Rise in Sea Level ... 77
Fig. 5.7. Severe Flash Floods in Europe ... 78
Fig. 5.8. Explosion in the Population of Jelly Fish ... 78
Fig. 5.9. Forest Fires in Australia ... 79
Fig. 5.10. A Satellite Picture of Fires raging across Australia ... 80
Fig. 5.11. Recurring Prolonged Droughts in India ... 81
Fig. 5.12. A Summary of Signs of Global Warming, NOAA ... 83
Fig. 6.1. Rock Striation, Mount Rainier National Park, Washington, USA ... 85
Fig. 6.2. Various Causes of Heating or Cooling of The Earth ... 86
Fig. 6.3. Ancient Air Trapped in Ice ... 87
Fig. 6.4. Climate of The Planet Earth for the Past 800,000 Years ... 89
Fig. 6.5. A News Item on Warming of Atmosphere Published in 1912 ... 90
Fig. 6.6. Temperature of The Earth and Concentration of Carbon Dioxide ... 91
Fig. 6.7. Global Greenhouse Gas Emission by Economic Sector for the Year 2010 ... 92
Fig. 6.8. Global Greenhouse Gas Emissions by Gas for the Year 2010 ... 93

Fig. 6.9. Concentration of Methane and Global Temperature 94
Fig. 6.10. Concentration of Methane 95
Fig. 7.1. Life Expectancy Rises with Availability of Power/Electricity 101
Fig. 7.2. The Gross Domestic Product Rises with Increased Availability of Power 101
Fig. 7.3. Human Development Index and Per Capita Power Consumption 102
Fig. 9.1. Formation of Coal 110
Fig. 9.2. A Coal Miner with a Canary 111
Fig. 9.3. Coal Miners 112
Fig. 9.4. Coal Wagon Trains 112
Fig. 9.5. A Coal Mine in Dhanbad 113
Fig. 9.6. Formation of Oil and Natural Gas 114
Fig. 9.7. Environmental Effects of Oil Spill 115
Fig. 9.8. Burning Gas Well 116
Fig. 9.9. A Forest After an Acid Rain 116
Fig. 9.10. Acid Rain Breaks Down Calcium Carbonate in Sandstone and Marble 117
Fig. 9.11. Semi-coke Heap, Spent Shale, and Open Pit Mines left after Shale Oil/Gas Extraction 119
Fig. 9.12. Fracking for Extraction of Oil or Gas 119
Fig. 9.13. Shale Gas Deposits in India 120
Fig. 9.14. Schematic Diagram of a Typical Coal Fired Thermal Power Plant 122
Fig. 9.15. Efficiency of use of Coal, Oil and Natural Gas 122
Fig. 9.16. Losing Energy at Every Step! 123
Fig. 9.17. Scheme for Coal Gasification 124
Fig. 9.18. Biomass 125
Fig. 9.19. Photosynthesis for Bioenergy 126
Fig. 9.20. Burning of Rice Stubbles 127
Fig. 9.21. Bagasse from Sugar Cane Industry 128
Fig. 9.22. Coconut Husk 129
Fig. 9.23. Forestation for the Rich and Poor Nations 130
Fig. 9.24. Biofuel 131
Fig. 9.25. Food vs. Fuel 132
Fig. 9.26. Biodiesel Production 133
Fig. 9.27. Carbon dioxide Emission per kWh of Power Produced 134
Fig. 9.28. Principle of Carbon Capture and Storage 136
Fig. 9.29. A Windmill in Kinderdijk, The Netherlands (Built in 1740) 137
Fig. 9.30. Windmills with Horizontal and Vertical Axes 138
Fig. 9.31. Paddy Fields and Wind Turbines in India 139
Fig. 9.32. Evolution of Global Wind Energy Installed Capacity in GW 140
Fig. 9.33. Snow Monkeys of Japan Taking Advantage of Hot Springs to Stay Warm 141
Fig. 9.34. Bison Keeping Warm During Winter Near a Steaming Hot Spring 141
Fig. 9.35. A Geyser in Haukadalur, Iceland 143
Fig. 9.36. The Principle of using Geothermal Energy 144

Fig. 9.37. Location of Most Active Geothermal Sites 145
Fig. 9.38. Global Installed Capacity of Geothermal Energy, July 2019 145
Fig. 9.39. Geothermal Map of India 146
Fig. 9.40. High Tide and Low Tide 147
Fig. 9.41. Tidal Turbine 148
Fig. 9.42. Generation of Electricity using a Tidal Barrage 149
Fig. 9.43. Wave Energy 151
Fig. 9.44. Asteroids Brought Water to Earth (Courtesy: NASA) 152
Fig. 9.45. The Water Cycle 153
Fig. 9.46. Principle of Hydroelectric Dam 154
Fig. 9.47. Floating Solar Power Station in Napa Valley, California 156
Fig. 9.48. Sardar Sarovar Dam, Gujarat 158
Fig. 9.49. Canal Top Solar Panels for Narmada Canal Net Work 159
Fig. 10.1. Sun 161
Fig. 10.2. Sun Rays Strike the Earth's Surface at an Angle 162
Fig. 10.3. Solar Radiation Spectrum at Sea Level After Going Through 1 Air Mass (AM1.0) 163
Fig. 10.4. Distribution of Solar Radiation across the World. The Area Covered by the Black Dots can meet the Energy Requirement of the Entire World! 164
Fig. 10.5. Lighting of Olympic Torch by a Parabolic Mirror 165
Fig. 10.6. Archimedes Burning Roman Ships Using a Parabolic Mirror, 212 BC 166
Fig. 10.7. A Typical Flat Solar Collector 166
Fig. 10.8. A Solar Thermal Heater 167
Fig. 10.9. Solar Bowl in Auroville, Puducherry 168
Fig. 10.10. Principle of Power Towers 169
Fig. 10.11. Ivanpah Solar Power Tower Facility 169
Fig. 10.12. Solar Furnace for 1 MW of Power 170
Fig. 10.13. The Solar Furnace at Odeillo in the Pyrenees-Orientales in France 171
Fig. 10.14. Parabolic Trough Reflector 172
Fig. 10.15. Parabolic Trough Solar Collectors at Solar Energy Generating Systems, California 173
Fig. 10.16. Molten Salt Storage Tanks under Construction at Solana Generating Station, USA to Provide 280 MWs of Power for 6 hours 174
Fig. 10.17. Principle of Fresnel Linear Reflecting Concentrators 175
Fig. 10.18. 100 MW Linear Fresnel Reflector Concentrating Solar Power Plant, Rajasthan 176
Fig. 10.19. Solar Dish Stirling Power System 177
Fig. 10.20. Main Parts of Alpha Stirling Engine (Zephyris, Wikipedia) 178
Fig. 10.21. Main Parts of Beta Stirling Engine (YK Times, Wikipedia) 178
Fig. 10.22. Main Parts of Gamma Stirling Engine 179
Fig. 10.23. Solar Cells Convert Solar Energy Directly to Electricity 180
Fig. 10.24. Metals, Semiconductors, and Insulators 181
Fig. 10.25. Basic Principle of a Solar Cell 182

Fig. 10.26. Silicon (Michel Bakni, Wikipedia) 183
Fig. 10.27. A n-type (left) and a p-type (right) Impurity in Silicon 183
Fig. 10.28. Working of a Solar Cell 184
Fig. 10.29. Spectral Losses in a Solar Cell. The Figure shows the Maximum Achievable Energy of a Silicon Solar Cell in Relation to the Sun Spectrum (AM1.5) 185
Fig. 10.30. Sunrays Absorbed by Copper Indium Gallium Selenide Solar Cells 186
Fig. 10.31. Efficiency of Solar Cells as a Function of Band Gap 187
Fig. 10.32. Dye Sensitised Solar Cells 189
Fig. 10.33. Tehachapi Energy Storage, California 192
Fig. 10.34. Pumped Up Hydro-electric Storage 193
Fig. 10.35. Reducing Cost of Silicon Photovoltaic Cells 194
Fig. 10.36. Global Growth of Solar Power 194
Fig. 10.37. Projections for Installed Capacity of Solar Power 195
Fig. 10.38. Pavagada Solar Park (2,050 MW) 196
Fig. 10.39. Cultivation of Betel Leaves in Shade 197
Fig. 10.40. Solar Panel Integrated Outdoor Mushroom Growing 197
Fig. 10.41. Duel use of Water: Solar Panel Washing and Irrigation 198
Fig. 10.42. Young Children Working in Cobalt Mines in Democratic Republic of Congo 199
Fig. 10.43. Requirement of Materials for Setting up Power Plants in tons/TWh 202
Fig. 10.44. Criticality of Availability of Materials during 2015–2025 203
Fig. 11.1. The Decay Chain of U-238 206
Fig. 11.2. The Nuclear Chart 207
Fig. 11.3. Absorption of Neutron by U-238 and Production of Pu-239 209
Fig. 11.4. Absorption of Neutrons by Th-232 and Production of U-233 210
Fig. 11.5. The Dynamics of Fission on Postage Stamps 211
Fig. 11.6. The Fission Chain Reaction 212
Fig. 11.7. Main Parts of a Nuclear Reactor 214
Fig. 11.8. Using Moderator for a Controlled Fission Chain Reaction for U-235 215
Fig. 11.9. Pressurised Light Water Reactor 216
Fig. 11.10. Neutron Economy in a Nuclear Reactor 217
Fig. 11.11. Nuclear Power Capacity of Different Countries in GWe (IAEA, 2020) 220
Fig. 11.12. Capacity Factor for Various Power Options 221
Fig. 11.13. Nuclear Energy Generation by Country (IAEA, 2019) 222
Fig. 11.14. World Energy Production by Source (2017) 222
Fig. 11.15. Growth of Nuclear Energy Production with Time 223
Fig. 11.16. Share of Nuclear Power in Total Energy Generation for Different Countries (IAEA, 2019) 223
Fig. 11.17. Pandit Jawahar Lal Nehru, Former Prime Minister of India Dedicating Apsara Reactor to Nation, January 20, 1957 226
Fig. 11.18. Uranium Resources of the World in kilo tons (Source: IAEA 2018) 227
Fig. 11.19. Three Stage Nuclear Power Programme of India 228
Fig. 11.20. A Typical Pressurised Heavy Water Reactor 229

Fig. 11.21. Sodium Cooled Fast Breeder Reactor (Wikipedia) 230
Fig. 11.22. Kalpakkam Mini Reactor (KAMINI), India with U-233 as Fuel 231
Fig. 11.23. Processes for Enrichment of Uranium 233
Fig. 11.24. Fast Breeder Test Reactor, Kalpakkam, India 235
Fig. 11.25. Kakrapar Nuclear Power Plant 236
Fig. 11.26. Molten Salt Reactor 238
Fig. 11.27. Scheme of Pebble Bed Reactor (Wikipedia) 239
Fig. 11.28. Graphite Ball for High Temperature Nuclear Reactor 240
Fig. 11.29. Structure of the Pebble 241
Fig. 12.1. A Part of the Digitised Sky with the Nearest Star Proxima Centauri (in red) 243
Fig. 12.2. Fusion of Protons Leading to Production of Energy in Sun and Other Stars 245
Fig. 12.3. Cross-section of Some Fusion Reactions, Including Those for DT Fusion 246
Fig. 12.4. Breeding of Tritium Using Fast Neutrons 247
Fig. 12.5. The Principle of Tokamak 248
Fig. 12.6. ITER ("The Way") Collaborators 249
Fig. 12.7. Visualization of ITER 249
Fig. 12.8. Indirect Drive for Fusion 250
Fig. 12.9. The Direct Drive 251
Fig. 13.1. The Patent of Carlo Rubia 255
Fig. 13.2. The Principle of Accelerator Driven Subcritical System 255
Fig. 13.3. The MYRRAH Experimental ADSS Reactor 256
Fig. 14.1. A Comparison of Mortality Rate for Various Sources of Energy 258
Fig. 14.2. Cumulative Years of Reactor Operation 258
Fig. 14.3. Reactor Vessel, India 259
Fig. 14.4. Containment Building for Pressurised Water Reactors 260
Fig. 14.5. Installation of Core Catcher at Rooppur Nuclear Power Plant, Bangladesh 261
Fig. 14.6. The Height of the Tsunami that struck the Fukushima Daiichi Station Approximately 50 minutes after the Earthquake 269
Fig. 14.7. Alpha, Beta, and Gamma Radiations 272
Fig. 14.8. Generation of Nuclear Waste 273
Fig. 14.9. Spent Fuel from a Light Water Reactor after using 1 ton of Uranium Enriched to 3.3% for Three Years 273
Fig. 14.10. Spent Fuel Pool 274
Fig. 14.11. Dry Cask Storage of Spent Fuel 275
Fig. 14.12. Open or Once-Through Fuel Cycle 275
Fig. 14.13. Closed Fuel Cycle 276
Fig. 14.14. Closed Fuel Cycle for Breeder Reactors 276
Fig. 15.1. Carbon-14 Dating 277
Fig. 15.2. I-131 is Used to Treat Hyperthyroidism 278
Fig. 15.3. Principle of Positron Emission Tomography 279

Fig. 15.4. A 99mTc-ECD (ethylene-cysteine-dimer) Brain SPECT Image demonstrating Bilateral Temporoparietal Hypoperfusion Typical of Alzheimer's Disease 280
Fig. 15.5. 100 Tons Per Day Sludge Treatment Plant in Ahmedabad 281
Fig. 15.6. Electron Beam Treatment of Wastewater 282
Fig. 15.7. High Yielding Groundnuts Developed by Bhabha Atomic Research Centre 283
Fig. 15.8. Embankments Made in the Catchment Areas to Revive Springs in Himalayas 284
Fig. 15.9. Direct and Indirect Ways to Radiation Damage of DNA 285
Fig. 15.10. Localized Deposition of Radiation Dose Using Particle Beams 286
Fig. 15.11. External Beam Radiotherapy and Brachytherapy 287
Fig. 16.1. Abundance of Elements in the Universe by Number of Atoms 289
Fig. 16.2. Space Shuttle Riding a Liquid Hydrogen and Liquid Oxygen Fuelled Rocket 291
Fig. 16.3. Schematic Diagram of a Typical Fuel Cell) 293
Fig. 16.4. A Hydrogen Fuel Cell Car 294
Fig. 16.5. The National Hydrogen Energy Road Map, Government of India, 2006 296
Fig. 16.6. Protype of a Train Running on Hydrogen 297
Fig. 16.7. Hy-4: First Passenger Aircraft Powered by a Hydrogen Fuel Cell 297
Fig. 16.8. Basic Scheme of Hydrolysis of Water using a PEM Electrolyzer 298
Fig. 16.9. Turquoise Hydrogen 300
Fig. 16.10. Hydrogen Economy for Isolated Hamlets 301
Fig. 17.1. A Vision for 100% Renewable Energy for 139 Countries 305
Fig. 17.2. Sheikh Saadi (1210–1292) in Rose Garden 307
Fig. 18.1. Mahatma Gandhi Arrives to Meet King-Emperor George-V (1931) 309

About the Authors

Prof. D. K. Srivastava (born 1952) is presently Homi Bhabha Chair Professor at National Institute of Advanced Studies, Bengaluru.

He graduated from the Allahabad University in 1970 and joined the Training School of the Bhabha Atomic Research Center, Mumbai. He started working at the Variable Energy Cyclotron Project of Bhabha Atomic Research Centre in 1971 and retired as Director and Distinguished Scientist at the Variable Energy Cyclotron Center, Kolkata in 2016. Later, he continued there as DAE Raja Ramanna Fellow until 2019.

He is a Fellow of National Academy of Sciences, India, and Indian National Science Academy. He was also conferred the Honorary Professorship of the Amity University in 2019.

He was awarded Indian Nuclear Society's award for "Outstanding Contribution to Teaching of Nuclear Sciences". Dr. Srivastava was given Outstanding Referee Award by American Physical Society (APS) in 2009. He was honored with Life-time Achievement Award by Chitkara University in 2012. He was also awarded Homi Bhabha Lecturer Award by Indian Physics Association and Institute of Physics UK in 2016. He has delivered Sir C. V. Raman Memorial Award Lecture at Calcutta University and Prof Abdus Salam Memorial Award Lecture at Jamia Milia Islamia, New Delhi.

He is serving as an Editorial Board Member of Pramana and Scientific Reports. He was also a Member of the Editorial Board of Physical Review C, during January 1, 2010 to December 31, 2012. He is President of Indian Physical Society since 2014.

Dr. Srivastava has worked at several prestigious labs and universities abroad, including KfK (now KIT) Karlsruhe, GSI Darmstadt, Lawrence Berkeley National Laboratory Berkeley, Brookhaven National Laboratory Upton New York, University of Minnesota, Duke University, McGill University, University of Cape Town, Bielefeld University, and University of Frankfurt, etc.

His research interests include Electromagnetic probes of quark gluon plasma (QGP), relativistic hydrodynamics, production, and propagation of charm quarks in QGP, transverse flow, etc. Earlier, he made imortant contributions in theoretical understanding of elastic, inelastic, and break-up reactions of light nuclei. Prof. Srivastava has published more than one hundred and fifty papers and delivered more than 400 talks, seminars, and colloquia. He is an author of two collections of short stories and more recently two books of children's literature in science.

At present he is working on energy, environment, and science outreach along with the continuation of his research on quark gluon plasma.

Prof. V. S. Ramamurthy (born 1942) is a well-known Indian nuclear scientist with a broad range of contributions from basic research to science administration. Prof. Ramamurthy started his career in Bhabha Atomic Research Centre, Mumbai in the year 1963. He made important research contributions in the areas of nuclear fission, medium energy heavy ion reactions, statistical and thermodynamic properties of nuclei and low energy accelerator applications. During the period 1995–2006, Prof. Ramamurthy was fully involved in science promotion in India as Secretary to the Government of India, Department of Science & Technology (DST), New Delhi. He was also the Chairman of the International Atomic Energy Agency (IAEA) Standing Advisory Group on Nuclear Applications for nearly a decade. Other important assignments held by him include Director, Institute of Physics, Bhubaneswar (1989–1995), DAE Homi Bhabha Chair Professor in the Inter University Accelerator Centre, New Delhi (2006–2010), Chairman, Recruitment and Assessment Board, Council of Scientific and Industrial Research, New Delhi (2006–2010) and Director, National Institute of Advanced Studies, Bengaluru (2009–2014). After retirement from service, Prof. Ramamurthy has also been actively involved in human resource development in all its aspects. He is currently Emeritus Professor, National

Institute of Advanced Studies in Bangalore. In recognition of his services to the growth of Science and Technology in the country, Prof. Ramamurthy was awarded the Padma Bhushan by the Government of India in 2005.

Chapter 1

Prologue

We are blessed to be living in a beautiful world, with its glorious sunrises and sunsets, its majestic animals, beautiful birds, varieties of insects, magnificent trees and plants, flowers with their divine charms and fragrances, *etc.* There is even a more complex world of single and multicellular organisms like bacteria and viruses beyond our vision capabilities but all the same very closely connected to our life cycles. More than 70% of the Earth's surface is covered by water. The underwater world supports its own distinctly different ecosystem with a variety of fishes, algae, plants, other marine animals, and corals, *etc.* About 10% of the Earth's surface is always covered with ice and again supports its own ecosystem. We have some form of life even up to a few kilometres under the surface of the Earth and up to two kilometres below the bottom of the sea with their own distinct life cycles. And together, they make an ecosystem where every species supports and depends on every other species for its survival. This ecosystem has taken a few billion years to evolve and has survived and recovered from several cataclysmic events.

The literature of the world is full of descriptions of the beauties of nature, painters have paid rich tributes to these and more recently, photographers have recorded these and made them available to everyone.

Recall one of the iconic descriptions of the beauty of morning by Janaki Vallabha Shastri (1916–2011):

"उदयति मिहिरो,विगलति तिमिरो
भुवनं कथमभिरामम्
प्रचरति चतुरो मधुकरनिकरो
गुंजति कथमविरामम्॥
विकसति कमलं,विलसति सलिलम्
पवनो वहति सलीलम्

दिशिदिशि धावति,कूजति नृत्यति
खगकुलमतिशयलोलम्॥

शिरसि तरूणां रविकिरणानाम्
खेलति रुचिररुणाभा
उपरि दलानाम् हिमकणिकानाम्
कापि हृदयहरशोभा॥"

"The Sun is rising; the darkness is being vanished. How enchanting is the world! The wise bumble bee has emerged, and you can hear its buzzing. The lotus is blooming, waves are playing on the surface of the water, and a gentle breeze is blowing. The birds are running, singing, and dancing everywhere. The sunrays are playing on the crowns of the trees and spreading the splendour of the Sun. The charm of the dewdrops glistening on the tips of blades of grass is stealing everyone's heart."

Music across the world is inspired by Nature; birds, rains, wind through trees, gurgling streams, and storms have all had their effects on musical compositions. Ludwig van Beethoven (1770–1827) loved Nature. One of his best loved works, Symphony No. 6, The Pastoral, is a testimony to it. Most of the symphony is a gentle meander through woods, fields, and by streams, taking in the smell and sounds it strolls along, including a storm that lashes over the land. In fact, he even wrote down the names of the birds; nightingale, quail, and cuckoo whose calls he included in the score.

The seven basic notes, swars (sound), of Indian classical music; sa, re, ga, ma, pa, dha, and ni, which replace the Western notes; do, re, mi, fa, sol, la, and si (or ti) are believed to have their origin in animal and bird sounds [1]. Thus,

- The first swar, sa, stands for Shadaja and represents the call of the peacock.
- The second swar, re, stands for Rishabha, and represents the sound of the bull.
- The third swar, ga, stands for Gandhar, and represents the sound of a goat or a sheep.
- The fourth swar, ma, stands for Madhyam, and is the call of the crane.
- The fifth swar, pa, stands for the sound of the koel, and is the short form of Pancham.

- The sixth swar, dha, stands for Dhaivat, and represents the sound of the frog.
- Finally, the seventh swar, ni, stands for Nishad, and represents the majestic call of the elephant.

These have been used to compose innumerable number of ragas and raginis, representing various moods, seasons, and times of the day and night. Raga Malhar, for example, portrays the life-giving rains and Raga Hansdhwani celebrates the call of swans.

The observation of various seasons, animals, birds, and natural events (Fig. 1.1) on the Earth gave rise to all the Sciences and man's quest for understanding the laws of Nature has continued to enrich his intellectual life and make his living more comfortable and fulfilling.

Fig. 1.1: The Sun Rising over a Desert [2]

Who is not enchanted by the enduring charm of the Rainbow Eucalyptus tree? Or, should one mention the immortal banyan trees — declaring to all "Quot Rami Tot Arbores" (as many branches, so many trees) and which have a capacity to cover an entire forest and which can give shelter to millions of birds, monkeys, and insects? Or maybe, we can recall the magnificent Giant Redwood Trees, having a girth of 30 metres, rising to 100 metres and a standing witness to the progress of man for more than 3,500 years? A Wisteria tree in Japan, a Gulmohar tree in India, a Rainbow Eucalyptus tree in Australia (Fig. 1.2), a Jacaranda tree in South Africa in full bloom, or fall colours in hills fill us with a divine bliss.

Fig. 1.2: A Rainbow Eucalyptus Tree in Australia [3]

The majesty of an African elephant on the move (Fig. 1.3), a cheetah in the pursuit of its prey, a Gir Lion surveying his "territory", a Royal Bengal Tiger looking at the world with its eyes burning like coal (Fig 1.4), an Indian bison standing proudly in its toned muscles of 1000 kilograms (Fig. 1.5), a swan moving most lyrically on a lake, a bird of paradise spreading its wings in Papua, New Guinea (Fig. 1.6), a monal — the most beautiful bird of Himalayas in its iridescent feathers, countless varieties of butterflies (Fig. 1.7) and grass-hopers, and a whale (Fig. 1.8) or a pod of dolphins jumping joyfully in the seas, flowers most beautiful in their vibrant colours, shapes, sizes and varieties (Fig. 1.9), and even the bacteria and human cells bring out the beauties and charm of the world. One lifetime, as it is said, is not enough to see all the charms of the world. And the world is beautiful!

Fig. 1.3: An African Elephant [4]

Fig. 1.4: A Royal Bengal Tiger [5]

Fig. 1.5: An Indian Bison, Jaldapara [6]

Fig. 1.6: A Bird of Paradise from Papua, New Guinea [7]

Fig. 1.7: A Ulysses Butterfly [8]

Fig. 1.8: A Humpback Whale Breaches Far in the Air above the Pacific Ocean [9]

Fig. 1.9: A Dahlia [10]

Of course, one of the most notable creations of Nature is man, with his vast and innumerable faculties — including sapience or faculty to be wise. William Shakespeare famously exclaimed (Hamlet, Act-2, Scene-2):

"What a piece of work is man,
How noble in reason,
How infinite in faculty,
In form and moving
How express and admirable,
In action how like an angel,
In apprehension how like a god,
The beauty of the world,
The paragon of animals!"

The Shwetashwar Upanishad called the humans, children of immortal bliss ("अमृतस्य पुत्राः").

Children, and for that matter babies of all living beings, are the living proof of vitality of life and present one of the most beautiful beings of this beautiful world. They fill us with wonderment and joy. Mother and child have remained a recurring theme in poetry, dramas, ballets, and paintings. Rabindranath Tagore (1861–1941) wrote about the beauty of children:

"The sleep that flits on baby's eyes -
does anybody know from where it comes?
Yes, there is a rumour that it has its dwelling
where, in the fairy village among shadows of the forest
dimly lit with glow-worms,
there hang two timid buds of enchantment.
From there it comes to kiss baby's eyes.

The smile that flickers on baby's lips when he sleeps -
does anybody know where it was born?
Yes, there is a rumour
that a young pale beam of a crescent moon touched
the edge of a vanishing autumn cloud,
and there the smile was first born
in the dream of a dew-washed morning -
the smile that flickers on baby's lips when he sleeps.

The sweet, soft freshness that blooms on baby's limbs -
does anybody know where it was hidden so long?
Yes, when the mother was a young girl it lay pervading her heart
in tender and silent mystery of love -
the sweet, soft freshness that has bloomed on baby's limbs."

-Gitanjali:61

Yet, it is not often realized that our world is limited to a thin sheet of atmosphere, a few kilometres thick, covering a near spherical planet called the Earth. The space vacuum beyond the atmosphere is not known to support life of any kind. As far as we know, Earth is the only planet harbouring life in its magnificent variety, since a few billion years. The vitality of life has survived repeated ice-ages, mega volcanic eruptions, huge tectonic plate movements, and even large asteroid impacts which wiped out the dinosaurs. Humans are a relatively recent species to inhabit the Earth. They are also the most influential.

It is also important to note that the lone star in our neighbourhood, the Sun, "powers" life on Earth in multiple ways. It "powers" the plant world and feeds the entire world. It provides the "power" for the oxygen factory in green leaves and sustains life. It "powers" the global water cycle, evaporates the water from the oceans and moves it as clouds and rain. It is this realization that evoked a reverence for the Sun in almost all civilizations. Suryasukta in Rigveda (1.115) and Aditya Hridayam Stotra in Valmiki Ramayana exalt the glory of the Sun. One of the most revered prayers in Rigveda, the Gayatri Mantra, places the Sun in an exalting position:

"भूर्भुवः स्वः तत्सवितुर्वरेन्यं । भर्गो देवस्य धीमहि, धीयो यो नः प्रचोदयात् ।।"

"*Let us meditate on that excellent glory of the divine vivifying Sun. May he enlighten our understandings.*" Rigveda (3.62.10) (Translation by Monier Monier-Williams)

Ancient Egyptian worshiped Sun [11] as the god Ra (Fig. 1.10)

The Choctaw Native Americans sing,

"The track of the Sun
across the sky
leaves its shining message,
Illuminating,
Strengthening,
Warming
us who are here,
Showing us we are not alone,
we are yet ALIVE!
And this fire...
Our fire...
Shall not die."

-Ref. [12]

Fig. 1.10: Ra, the Sun god of ancient Egyptians [11]

Fig. 1.11: The Inheritors of the Earth [13]

We realize that the history of mankind is the history of his search for resources to meet his growing needs. It is essential that his search does not destroy this wonderful tapestry of life and the ecosystem of the Earth.

Let us never forget the most apt remark, "We have not inherited this Earth from our forefathers; we have borrowed it from our children." It is our bounden duty to preserve this beautiful Earth with its wondrous variety of life for them (Fig. 1.11).

Chapter 2

The World is Fragile

The beautiful world that we described in the previous Chapter is also a fragile world. Nature has, over the last millions of years, evolved a web of life which maintains a very delicately balanced biodiversity.

Let us start with a story of hope and insight. When wolves were hunted into oblivion in the Yellow Stone Park in the USA in 1920s, the population of elk proliferated. It led to destruction of grazing land. Then the elks started eating tree saplings and smaller plants. The loss of vegetation led to erosion of land and drying up of the resident river. This led to destruction of habitat for bear, otters and fish, whose population decreased.

Wolves were reintroduced there in 1990. Now the elk population is under control, grasslands and forest vegetation have grown back, land erosion has stopped, the river is regenerated, and fish and otter colonies are back and bear feeding on abundant berries have multiplied. The regenerated forest now reverberates with bird songs, sounds of gurgling streams, and howls of wolves on a moonlit night.

Ecosystems are also destroyed by natural disasters like earthquakes and tsunamis, forest fires, or volcanic eruptions. However, Nature has evolved strategies to restore them once the damaging event has passed. An asteroid impact, 66 million years ago, near Chicxulub in the Gulf of Mexico led to an abrupt end of most dinosaurs along with about three-quarters of all life on Earth. Life on Earth has suffered five mass extinctions of biodiversity in its long history, caused by massive volcanic eruptions, deep ice ages, meteorite impacts, and clashing of continents. And it has recovered from each one of these natural disasters.

Nature has also devised its own renewal strategies. For example, there are several pine trees which do not even germinate unless their parent tree has burned down in a wildfire. In view of this, biodiversity is seen as the knowledge acquired by evolving species over millions of years about how to survive through the vastly varying environmental conditions that Earth has experienced.

Human beings have been continuously altering the environment through agriculture, logging, mining, construction, pollution, *etc.* When we destroy the biodiversity, the ecosystem is unable to provide the services it has been providing since the beginning of life on Earth. Our survival depends on biodiversity, yet we continue with increasing biodiversity destruction.

A recent report prepared by the United Nations Intergovernmental Science-Policy Platform on Biodiversity and Ecosystem Services (2019) concludes that "The health of ecosystems on which we and all other species depend is deteriorating more rapidly than ever. We are eroding the very foundations of our economies, livelihoods, food security, health and quality of life worldwide." It further noted that, "Ecosystems, species, wild populations, local varieties and breeds of domesticated plants and animals are shrinking, deteriorating, or vanishing. The essential interconnected web of life on Earth is getting smaller and increasingly frayed. The loss is a direct result of human activity and constitutes a direct threat to human well-being in all regions of the world."

It is estimated that around 1 million animal and plant species are now threatened with extinction, many within decades, more than ever before in human history. The abundance of native species is estimated to have decreased by at least 20%, since 1900. About one-third of reef-forming corals and marine animals and two-fifths of amphibian species are threatened. About 700 vertebrate species since 16th century have been driven to extinction. Many domesticated breeds of animals are extinct, and 1000 more breeds may go extinct by 2020s.

Here we shall concentrate on human intervention, which has brought about an unprecedented biodiversity loss in recent times; especially following industrialization — with discoveries of coal powered steam engines, internal combustion engines, and electricity — all using hydrocarbons. There are also suggestions that human activities started seriously affecting the biodiversity with the advent of agriculture. These activities can be summarised as: deforestation, introduction of invasive species, pollution, climate change, and over-exploitation. Conflict is avoidable, yet it continues causing grievous harm to biodiversity.

Human conflicts and wars have also caused considerable biodiversity loss in the past. Ancient Romans used to sow salt in the agricultural lands of their foes to make these barren. During the Vietnam War, US forces sprayed herbicides like Agent Orange on forests and mangrove swamps to eliminate covers for guerrilla soldiers. Some of those regions are not likely to recover for decades. A series of armed conflicts in the Democratic Republic of Congo since 1990s have had a devastating effect on wildlife populations which became a source of bush meat for combatants, civilians struggling for survival, and commercial traders. Species like antelopes, monkeys, rodents, apes, and elephants have suffered enormously on this account.

Saddam Hussein's troops drained the Mesopotamian marshes, in 1990s. These were largest wetland ecosystems in Middle East and situated at the confluence of the Tigris and Euphrates rivers. The marshes were reduced to less than 10 per cent of their original size. The landscape was transformed into a desert with salt crusts. Again in 2017, militants set ablaze oil wells in the city of Mosul which released a toxic cocktail of chemicals into the air, water, and land.

Decades of conflicts in Afghanistan have destroyed more than 50% of the forests, making the land vulnerable to floods, avalanches, and landslides.

Deforestation

With the rise in human population, wild areas are razed for farmland, cattle, soy, palm oil, timber, and leather. More than 550,000 hectares of dense forests in Assam, Darjeeling, and South India in mid-nineteenth century were cleared for plantation of tea. It is estimated that 100 million hectares of forests were cutdown in the tropics during 1980–2000 for cattle ranching in Latin America and palm oil plantations in Southeast Asia. This brings animals and humans into conflict, which the animals invariably lose. Poaching and unsustainable hunting for food is leading to likely extinction of more than 300 mammal species from chimpanzees to hippos to bats.

Another worrisome consequence of deforestation has now emerged. We know that malaria is preventable and curable. Yet, it continues to affect people of more than 90 countries. In 2015 alone, there were 212 million cases of malaria and about 430,000 deaths. More than 90% of these cases were in Sub-Saharan Africa. It has been found that deforestation which leads to standing water, increasing temperatures and more sunlight, is quite favourable to most malaria-carrying species.

Invasive Flora

Flora and fauna native to a place have taken millions of years to adopt to their surroundings and to be in harmony with it. Birds rarely eat fruits from or make nests on non-native trees. Leaves of non-native trees are only rarely eaten by local animals. Eucalyptus trees (natives of Australia) were planted all over India in large numbers in late seventies as they grow very fast. But their leaves are not eaten by Indian animals. In fact, no animal other than koala bears (only found in Australia), eats them.

Fig. 2.1: Congress Grass in India [14]

Congress Grass (Parthenium hysterophorus) arrived in India with the PL-480 wheat. It has spread across India as a weed and causes severe harm to people and animals. It produces chemicals that suppress crop and pasture plants, and allergens that affect humans and livestock. It also frequently causes pollen allergies.

Angreji Babool (Prosopis juliflora), a native of Mexico, was introduced in India to provide firewood by the English and Maharaja of Jodhpur in 1940s. It is now an invasive weed in many nations. It is hard and expensive to remove as the plant can regenerate from the roots. Its water uptake is very high, which leads to elimination of native plants and even grass, causing land erosion. Its aggressive growth denies native plants water and sunlight, without providing food for native animals and cattle.

Fig. 2.2: Angreji Babool in India [15]

Fig. 2.3: Water Hyacinth [16]

Water hyacinth was introduced in India towards the end of 18th century by Lady Hastings, wife of its first British Governor-General. Now it has occupied all rivers, lakes, and ponds in entire South Asia, Africa, and Oceania. Its plants duplicate within nine days and very quickly cover lakes and ponds entirely. It affects water flow and blocks sunlight from reaching native aquatic plants which die. The decay depletes dissolved oxygen in water, killing fish and turtles. The plants provide a prime habitat for mosquitos and several other parasitic worms. They provide shelter to mosquitos, which leads to occurrence of malaria. Hydrilla plants from Sri Lanka or Africa have had a similar success in choking water bodies in USA and eliminating local plants.

Invasive Fauna

Just 24 wild rabbits brought from England in 1859 were introduced in Australia, for a sport of hunting. There were no predators for them in Australia, and their number rose to millions within 30 years. Vast stretches of fertile land with plenty of vegetation changed into dry areas, with no plants, little water, and drought.

Fig. 2.4: The Erosion of a Gully in South Australia Caused by Overgrazing by Rabbits [17]

The authorities encouraged people to hunt them and every year millions were hunted and yet they continued to multiply. To control their population, foxes were introduced. But they found plenty of preys — much easier to hunt (like land birds and rodents) and left the rabbits alone. Infecting the rabbits with a virus was tried, which killed about 99.8% of them, but the remaining 0.2% of them adopted to it and multiplied again!

Nile Perch was introduced in Lake Victoria for improved commercial and sports fishing in 1950s and became extremely prized as it grows to a length of 2 meters and attains a weight of up to 200 kg. However, soon it eliminated several hundred species of local fish including cichlids. As their food supply reduced, their numbers reduced too, but then they adopted to eating small shrimps and minnows.

Fig. 2.5: Nile Perch Introduced in Lake Victoria [18]

In India, Red Belly Piranha (from South Africa), African Cat Fish, Amazon Sailfin Catfish, Arapaima, Tilapia, Alligator Gar, Pangasius (from Malaysia), Pacu, and many other exotic non-native fish, which were perhaps being farmed illegally in dams or in private aquaria have escaped into rivers, dams, lakes and ponds. They have started eliminating the local fish.

Myna, a bird common across Indian subcontinent, was introduced in Australia to control insects in 1860s. By now, these are declared as pests. Similarly, bulbuls escaping from pet shops and cages have proliferated in New Zeeland and reached a stage where they are declared as pest, as they do extensive damage to fruits.

Fig. 2.6: Myna in Sydney [19]

Fig. 2.7: A Common Sparrow

When the Sparrows were smashed

During the movement called Great Leap Forward (1958–1962) in China, Mao Zedong gave a call to Smash Sparrows — calling them "public animals of capitalism", under the belief that they were eating huge amounts of food grains and fruits.

Millions of people from the countryside destroyed their nests, broke their eggs, and killed their chicks. They made loud noise by beating pots and the birds — exhausted and unable to rest, dropped dead. In a campaign involving up to 300,000 students, officials, and army personnel slaughtered thousands of sparrows in a day in Beijing.

Many of the birds took shelter on trees inside embassies of foreign countries in Beijing. The Embassy of Poland did not allow the pursuers to enter its premises. They surrounded it for two days and pots and drums were continuously beaten. The Polish authorities had to use shovels to remove the dead birds. In the absence of natural predators, the ecological balance was upset, insects destroyed the crops and caused the Great Chinese Famine which killed 15 to 45 million people.

Hunting and Poaching

Babur (1483–1530), the founder of Mughal Empire in India, hunted rhinoceros near Peshawar. Now they are confined to just a small corner of Assam and North East of India. They are killed in the mistaken belief that powder from their nose-horn acts as an aphrodisiac. African rhinoceros are facing extinction for the same reason. Elephants are killed as they come in conflict with humans and for their tusks. Many tigers were hunted to satisfy the ego of our British rulers and Maharajas. There is a strong demand for tiger parts in traditional Chinese medicine, and for their skins to be used as home décor. In Sunderbans, the shrinking forests are constantly bringing them in conflict with humans.

Dodo was a flightless bird which was first recorded by Dutch sailors in 1598 in Mauritius. By 1662, the last dodo was extinct due to hunting and introduction of invasive predators, like dogs, in the island. There are suggestions that a tree called tambalacoque or dodo tree was almost facing extinction, as according to one theory, its seeds had to pass through the digestive tracks of dodo birds to germinate. Fortunately, a large turtle has been found now, which provides this service.

Fig. 2.8: Babur Hunting Rhinoceros Near Peshawar [20]

Fig. 2.9: Maharajah Ramanuj Pratap Singh Deo with the Last Three Cheetahs, 1948 [21]

Cheetahs were widespread in India. They could easily be trapped, tamed, and trained to assist in hunting. Historical records show that the Moghul Emperor Akbar (1542–1605) had more than 7000 cheetahs in his collection. Cheetahs, like elephants, do not breed in captivity. In 1948, soon after the Independence, Maharaja Ramanuj Pratap Singh Deo of Koria, Chhattisgarh of Sarguja State, shot and killed the last three cheetahs in the wild in India. They were declared as extinct in 1952. The gentleman also held the dark honour of having killed 1200–1700 tigers.

During the Mughul, feudal, and the colonial times in British India, hunting was considered a regal sport that showcased the royalty, machismo, power, and wealth of the rulers. This resulted in large scale killing of tigers. There are suggestions that the slaughter was legitimized by vilifying the tigers, casting them as terrible, bloodthirsty beasts with an unquenchable desire for human flesh.

Over 80,000 tigers are believed to have been slaughtered in 50 years from 1875 to 1925. The killing escalated after 1947 as Independence ushered in a hunting free-for-all. In 1969, Mrs. Indira Gandhi initiated steps to protect tigers, whose numbers had reduced to about 1800 by 1971. At least 100,000 tigers were believed to roam our forests in 1900. Project Tiger started in 1973 and the tiger population rose by 4000 by 1984 when she was assassinated. It is one of the most successful preservation projects in the World.

Lions were widespread almost across the world. They were extinct in Greece by 100 BC. Now the only place outside Africa, where they are found, is in Gujarat where they were initially protected by Nawab of Junagadh and their number is about 700, now.

In the 16th century North America had 25–30 million American bison. They were hunted down to about 100 by 1880s. It is estimated that up to 100 African elephants are killed each day for their tusks.

Till mid-nineteenth century, billions of passenger pigeons filled the skies of North America. In fact, their flocks used to be described as “avian clouds”. The last passenger pigeon, lovingly called Martha was found dead on September 1, 1914. These birds went extinct as they were widely hunted and the trees on whose seeds (masts) they survived were cleared for farming.

More than 99.9% of vultures in India are believed to have died because of a banned drug, diclofenac, used by farmers to ease inflammatory pain in cattle.

Fig. 2.10: Shooting of Passenger Pigeons [22]

Of Bees, Butterflies, and Elephants

Bees are better known pollinators. Butterflies, beetles, moths, birds, and bats also do their fair share of pollinating fruits and crops. More than 90 percent of all plants need a pollinator to distribute pollen to set fruit and seeds.

In fact, many moths and beetles are specific to some plants, and their extinction is bound to lead to the extinction of their co-existential partner.

There are estimates that 250,000 to 500,000 species of insects have died out since industrial revolution. There are also estimates that bee population may have decreased by more than 40%. The loss of bees put up to US$ 577 billion in annual global crops at risk.

At several places in China, local pollinators, bees, and butterflies, have died out due to excessive use of pesticides, leaving farmers to pollinate their crops by hand or to hire bees. The wide-spread use of glyphosate-based herbicides is likely to trigger loss of biodiversity, making ecosystems more vulnerable to pollution and climate change. These have also seeped into waterbodies and soil, putting the life of animals and people to risk.

The demand of frog-legs in the export market, led to a sustained slaughter of adult frogs from paddy fields in West Bengal. This led to increase in number of insects and decrease in number of larger predators, thus disturbing the ecosystem. The farmers reacted by using more insecticides, whose residues poisoned the soil and water.

Fig. 2.11: Bees as Pollinators [23]

Many trees depend on animals and birds for their propagation, as they eat their fruits and spread it far and wide. This is necessary to remove competition for space, nutrients, water, and light and to protect their species from pests and disease which could lead to their extinction if the trees of a type crowded at one place.

Fig. 2.12: Hand Pollination of Apple Flowers in China [24]

Elephants and monkeys in Asia, Africa, and Amazon help in dispersal of seeds, some of which must pass through their digestive tracks to germinate. Squirrels tend to collect acorns and burry them at various places to be eaten later. Acorn propagates as the squirrels occasionally forget the burial points.

Pollution

Pollution is one of the biggest threats facing mankind, which is almost as severe as the climate change. We have polluted the soil — from where we get our food, the water — which we drink, and the air — which we breathe.

The indiscriminate use of pesticides is killing the useful bacteria and earthworms in the soil. The ecosystem, on land and in the water, depends heavily upon the activity of bacteria, which work ceaselessly to cycle nutrients like carbon and nitrogen. The glyphosate-based herbicides can trigger loss of biodiversity, making ecosystems more vulnerable to pollution and climate change. Increasing amounts of glyphosate is leaching into environment and contaminating water bodies.

Untreated effluents from industries like leather, textiles, and sugar mills release toxic chemicals into waterbodies. There are persistent reports of these harmful chemicals polluting underground sources of water, which are any way drying up due to excessive over-exploitation.

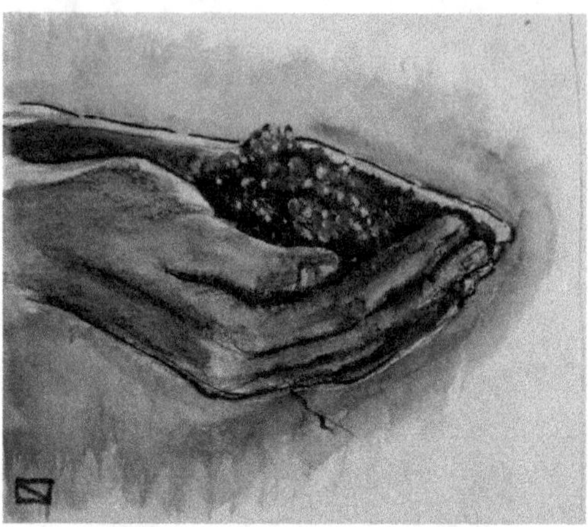

Fig. 2.13: A Handful of Soil May Contain Several Billion Bacteria [25]

It is estimated that 300–400 million tons of heavy metals, solvents, toxic sludge and other wastes from industrial facilities are dumped annually into the world's waters, and fertilizers entering coastal ecosystems have produced more than 400 'dead zones' in the oceans, totalling more than 245,000 sq. km.

Release of detergents containing phosphate salts is a big concern across the world. Phosphates inhibit the biodegradation of organic substances. They also cause the water bodies to get choked with algae and plants. The excessive richness of nutrients in a lake or other bodies of water, frequently due to run-off from the land (due to excessive use of fertilizers), causes a dense growth of plant life. This process is called eutrophication and deprives the water of available oxygen, causing the death of other organisms.

Our current pattern of consumption also produces an enormous amount of waste which Nature is unable to handle. The world generates about two billion tons of municipal waste annually. Closer home, metropolitan cities like Delhi or Mumbai produce up to 10 thousand tons each of municipal waste daily. The waste continues to mount, causing untold miseries to the neighbourhoods where these are dumped in improvised landfills, ditches, ponds, riverbanks, seashores, valleys, slopes, railway tracks and just roadsides. Every day, in most cities, we witness the heart-breaking sights of young children and grown-up people rummaging through the dumps for anything of value. The toxic chemicals from these landfills are poisoning the underground water system. From time to time, these dumps catch fire due to formation of methane gas which along with the fumes of other toxic substances poisons the atmosphere.

Fig. 2.14: Plastic Pollution in Yangtze River near Shanghai [26]

The other menace arises from use of plastics. Mass production of plastics began about six decades ago. We produce about 300 million tons of plastic every year now. By now about 8.5 billion tons of plastic have been produced. This is more than one ton per head! Most of the plastic is used for making disposable products that end up as trash. Only about 10–20% of it has either been recycled or incinerated. The rest has been dumped around in landfills or just garbage dumps. Up to 8 million tons of this plastic find its way into oceans every year. As plastic may take up to 400 years to decompose, the hazard is obvious.

Plastic bags and plastic products are regularly being recovered from the guts of fish, animals, and even birds, which are also often getting asphyxiated in the discarded plastic bags. In a very painful, dastardly, and diabolical episode, hundreds of thousands of tons of this plastic are dumped by western nations in countries like China, India, Vietnam, Indonesia, Malaysia and African countries — often by deliberately mislabelling them as suitable for recycling. China and India have recently banned the entry of this plastics into their countries — but at least in India a large amount of plastic waste is still entering surreptitiously. Open air burning, often resorted to, releases very toxic fumes and particulate matter into the atmosphere.

It has now been realized that several cosmetics, special cleaning materials, synthetic clothes, ropes, nets, and all kinds of plastic objects release microplastics (pieces less than 5 mm in length) which are entering our food chain. The fullest impact of these has not yet been studied in detail.

Over-exploitation

Over-exploitation is a serious issue, especially when it comes to fishing. In 1990s, fishing in East Coast of Canada collapsed due to overfishing. It was blamed on seals, which were slaughtered in large numbers. Two thirds of the Earth are covered by oceans. Seafood provides protein to more than 2.5 billion people. However, rampant overfishing has caused the catches to fall steadily. Now more than half the ocean is industrially fished. In 2015, 33% of marine fish stocks were being harvested at unsustainable levels.

The air pollution and its effects are only too well known and will be discussed in more detail later, along with the climate change. We only add that up to 8 million people are estimated to meet a premature death every year due to pollution caused by burning of fossil fuel.

A famous Native American (Cree) saying comes to our mind, which presciently declares, "When the last tree has been cut down, the last fish caught, the last river poisoned, only then will we realize that one cannot eat money."

Only a very concerted and sustained global, national, and local effort along with suitable laws can maintain the delicate balance of the biodiversity and the ecosystem of our Earth.

Chapter 3

A Little More About Mother Earth

Fig. 3.1: Earth Seen from Space [27]

It is often said that the only certainty on the face of the Earth is the rising and setting of the Sun every morning and evening. What do we see when we look towards the sky in the night? There is the Moon, waxing and waning and moving slowly across the sky and the stars always engaged in a heavenly dance. Seen from space, the planet Earth presents one of the most beautiful sights, with its blue oceans, emerald seas, different continents, ice-capped poles, swirling clouds, snow covered lofty mountains, mighty rivers, verdant forests, and vast deserts.

How was the Earth formed? How were the Sun, the Moon and the stars formed? What are we made of? What are the stars? What leads to seasons? How did life originate? Why is life so diverse and varied? The search for answers to questions like these has always been intriguing the human mind and gave rise to the evolution of human civilization.

Fig. 3.2: Basic Building Blocks According to Ancient Greeks [28]

For a long time, it was believed that the material universe is made up of the four basic elements — fire, earth, air, and water, which had the properties of being hot, cold, dry, and wet. Thus, for example, a rock was believed to have more of earth, with moderate amounts of fire, air, and water. A rabbit, on the other hand, had more of fire and water and a moderate amount of air and earth.

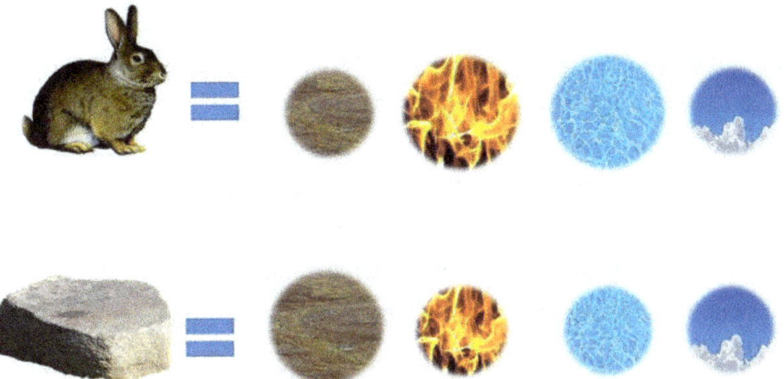

Fig. 3.3: A Rabbit and A Rock [28]

A lone conjecture by Democritus (460 BC–360 BC) in 390 BC that there could exist the smallest indivisible part of matter, the Atomos (indivisible), had to wait a few centuries to find common acceptance. He asked, "Could matter be divided into smaller and smaller pieces forever or was there a limit to the number of times a piece of matter could be divided?" He called the smallest piece as "Atomos", which could be different for different matters.

Fig. 3.4: Atomos of the Fur of a Lion and a Rock According to Democritus [28]

What led to this long delay? Two eminent philosophers of that time, Aristotle and Plato had a more respected (and ultimately wrong!) theory. Aristotle and Plato favoured the earth, fire, air, and water approach to the nature of matter. Their ideas held sway because of their eminence as philosophers. The idea of Atomos was buried (had to wait) for approximately 2000 years!

It was in early 1800's that the English chemist, John Dalton, performed several experiments that eventually led to the acceptance of the idea of atoms.

He deduced that:

- All elements are composed of atoms.
- Atoms are indivisible and indestructible particles.
- Atoms of the same element are exactly alike.
- Atoms of different elements are different.
- Compounds are formed by the joining of atoms of two or more elements. It does not involve the destruction of the atom.
- When elements react to form compounds, they react in definite ratios of whole numbers.

This theory laid the foundations of modern chemistry and ultimately led to the realization of periodic behaviour of elements and then their arrangements in Dmitri Mendeleev's Periodic Table in 1869. The crowning glory of Mendeleev's realization was that that the chemical properties of the elements repeat themselves according to a regular pattern, that is, the properties are periodic.

Group→ ↓Period	1	2	3		4	5	6	7	8	9	10	11	12	13	14	15	16	17	18
1	1 H																		2 He
2	3 Li	4 Be												5 B	6 C	7 N	8 O	9 F	10 Ne
3	11 Na	12 Mg												13 Al	14 Si	15 P	16 S	17 Cl	18 Ar
4	19 K	20 Ca	21 Sc		22 Ti	23 V	24 Cr	25 Mn	26 Fe	27 Co	28 Ni	29 Cu	30 Zn	31 Ga	32 Ge	33 As	34 Se	35 Br	36 Kr
5	37 Rb	38 Sr	39 Y		40 Zr	41 Nb	42 Mo	43 Tc	44 Ru	45 Rh	46 Pd	47 Ag	48 Cd	49 In	50 Sn	51 Sb	52 Te	53 I	54 Xe
6	55 Cs	56 Ba	57 La	*	72 Hf	73 Ta	74 W	75 Re	76 Os	77 Ir	78 Pt	79 Au	80 Hg	81 Tl	82 Pb	83 Bi	84 Po	85 At	86 Rn
7	87 Fr	88 Ra	89 Ac	**	104 Rf	105 Db	106 Sg	107 Bh	108 Hs	109 Mt	110 Ds	111 Rg	112 Cn	113 Nh	114 Fl	115 Mc	116 Lv	117 Ts	118 Og
				*	58 Ce	59 Pr	60 Nd	61 Pm	62 Sm	63 Eu	64 Gd	65 Tb	66 Dy	67 Ho	68 Er	69 Tm	70 Yb	71 Lu	
				**	90 Th	91 Pa	92 U	93 Np	94 Pu	95 Am	96 Cm	97 Bk	98 Cf	99 Es	100 Fm	101 Md	102 No	103 Lr	

Fig. 3.5: Dmitri Mendeleev's Periodic Table [29]

Fig. 3.6: Dmitri Mendeleev and his Periodic Table [30]

Only 63 elements were known during the time of Mendeleev. Now we know 118 elements. They are represented in the table by their symbols, e.g., C for carbon and Au for gold (Aurum in Latin). The symbols additionally have two numbers associated with each element. The first of these numbers is called the atomic number which, we now know, stands for the number of protons in the nucleus of the atom (see later). It is equal to the number of electrons — which have equal and opposite electric charge, and which makes the atoms neutral. The chemical properties and how a given atom combines with other atoms to form a compound are decided by this number and arrangement of the electrons.

The second number associated with each element is called the atomic mass. It is given by the sum of number of neutrons and protons in the nucleus. Neutrons, as the name suggests, have no electric charge. Hydrogen, for example, has an atomic mass close to 1 and Uranium has an atomic mass close to 238, implying that an atom of Uranium is about 238 times heavier than a hydrogen atom. The nuclei of some elements have a given number of protons but differing number of neutrons. These are called isotopes (as they have the same place in the periodic table). The relative abundances of the isotopes of a given element may vary considerably and are decided by their nuclear structure and binding energy (see later).

Protons and neutrons were unknown at the time of Mendeleev and he arranged the elements according to increasing mass. Emphasizing on their chemical properties, he left spaces for elements he believed would be discovered. Based on the periodicity, he predicted their properties. Gallium (1875), scandium (1879) and germanium (1886) were discovered with those very properties, thus firmly establishing his Periodic Table. It may interest the readers to know that despite this path-breaking discovery, which ultimately paved the way for a clear insight into the atomic structure of elements, Mendeleev was not awarded Nobel Prize. UNESCO designated 2019 as the International Year of the Periodic Table to celebrate 150 years of its discovery.

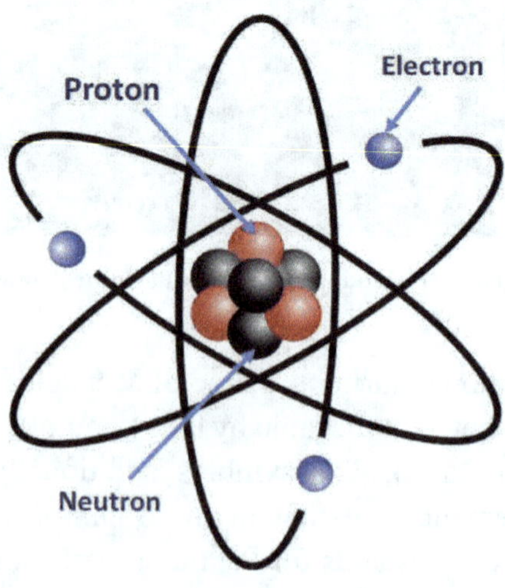

Fig. 3.7: Rutherford Atom [31]

Decades of brilliant and pioneering theoretical and experimental research by hundreds of scientists in the early twentieth century has now revealed that an atom has a small positively charged nucleus surrounded by an electron cloud with electrons occupying well-defined quantum states. The discovery of the neutron indicated earlier, completed the picture.

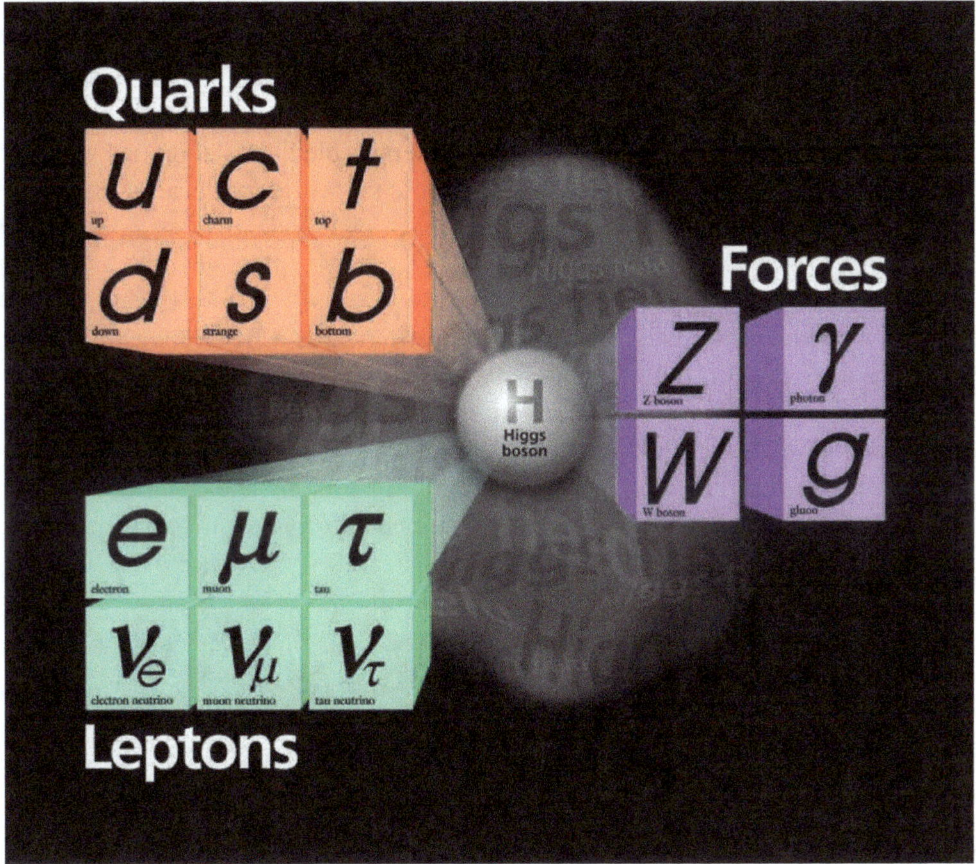

Fig. 3.8: The Basic Building Blocks of Matter [32]

It is now known that the protons and neutrons are themselves composite particles and are made of fractionally charged particles called quarks and their antiparticles. Gluons mediate the strong interaction between quarks. This interaction is described by a theory called Quantum Chromo Dynamics. Once we include electrons and neutrinos (a weakly interacting and extremely light neutral particle), a pattern in charge appears. The model has been so successful that it is called the Standard Model of particle physics. We add that photons mediate electromagnetic interaction between charged particles and W and Z bosons mediate weak interaction. Weak interaction is responsible for decay of a neutron into a proton (+ an electron and an anti-neutrino). The Standard Model does not deal with gravitational interaction. There was still one piece in the zig-saw puzzle missing. Higgs boson, predicted to provide masses to all the fundamental particles and postulated by Robert Brout, François Englert and Peter Higgs in 1964, was finally discovered at CERN, Geneva in 2012, completing the Standard Model.

The unravelling of the atomic and sub-atomic structures of the material universe within a few decades is one of the most outstanding and exciting scientific developments of the twentieth century — made possible by collaborative work of thousands of scientists and engineers from across the world for decades.

Formation of the Earth

It is now generally accepted that the Universe started as a Big Bang. In the first moments after the Big Bang, which is believed to have taken place around 13.8 billion years ago, the universe was extremely hot and dense. It expanded and cooled and gave rise to building blocks of matter — quarks and electrons. A few microseconds later quarks came together to form protons and neutrons. A few minutes later these protons and neutrons combined to from helium and then a trace of lithium and beryllium nuclei.

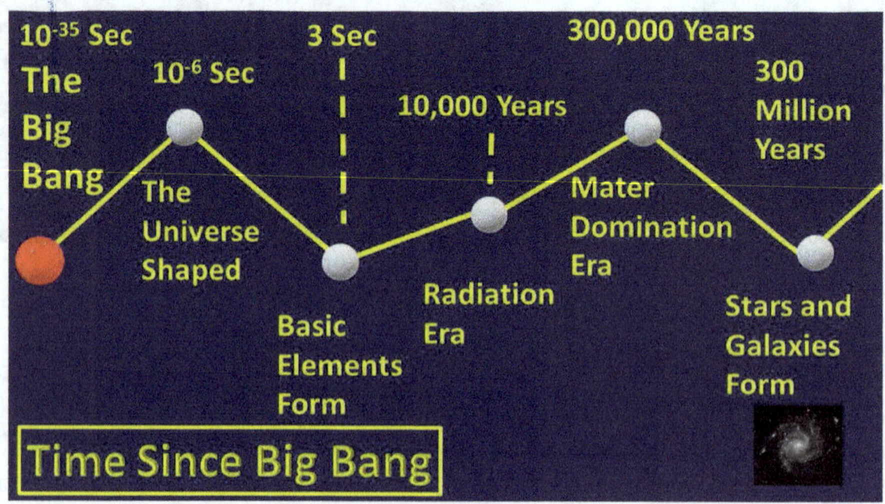

Fig. 3.9: History of the Early Universe [33]

It took about 380,000 years for electrons to be trapped in orbits around nuclei, forming the first atoms. These were mainly helium and hydrogen, which are still by far the most abundant elements in the universe. This also led to decoupling of photons from the medium which fill the Universe with Cosmological Microwave Background Radiation today and whose small anisotropy retains the memory of initial fluctuations in the density of the matter in the Universe, which is responsible for the structures in the Universe.

Finally, about 1.6 million years later, gravity led to formation of stars and galaxies from the clouds of these gases. These were the first generation of stars. The elements up to iron in the periodic table were cooked through nuclear fusion inside stars and heavier elements through absorption of neutrons and beta-decay upon explosion of the stars as supernovae. These filled the space with dust and gas of these elements and formed nebulae.

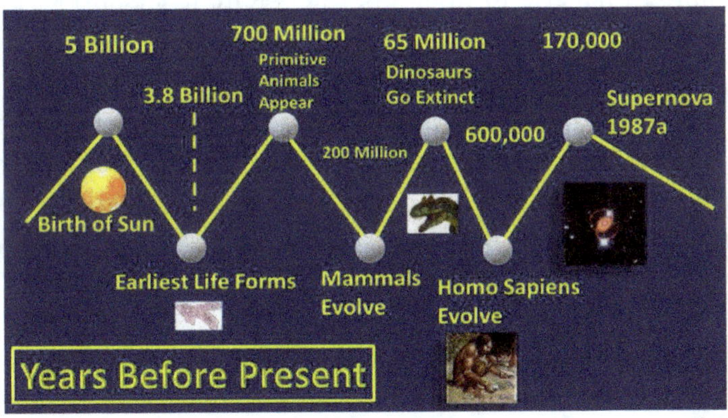

Fig. 3.10: Evolution of Life on Earth [33]

About 4.5 billion years ago, the area occupied by our solar system was originally occupied by one such nebula of interstellar dust and gas. Another supernova explosion in the neighbourhood, created a shock wave which flattened the cloud and set it spinning. Gravity pulled more and more of the matter at the centre and led to a mass of matter whose core was under high pressure and high temperature- so high that, that it triggered nuclear reactions. Nuclei of hydrogen fused to form helium and our Sun was born. These reactions in the Sun convert about 4.35 tons of matter into energy per second at the rate of 3.85×10^{26} Watts per second. The Sun appropriated about 99.8% of the total mass of the cloud for itself. The remaining matter formed planets and moons, and the asteroid belt of smaller and larger pieces of matter. It also gave birth to several dwarf planets and comets.

The heat nearer to the Sun was so large that only rocky materials could withstand it and we have Mercury, Venus, Earth, and Mars as rocky planets. The strong solar wind also blew the light gases like hydrogen and helium away from these. The planets further away are made up of mostly hydrogen and helium gas — *viz.*, Jupiter and Saturn, or ice — *viz.*, Uranus and Neptune. All planets, barring Mercury, and Venus, have moons. The solar system has more than 150 moons.

Mercury and Mars have weaker gravity (about 38–40% of that of Earth) and very thin atmospheres. This, coupled with their proximity to Sun, leads to vast variations of their temperature as day changes to night. For example, the temperature of Mercury during the day is about 425 degree Celsius and during the night it drops to 170 degree Celsius below zero. The day temperature of Mars is about 20 degree Celsius, but the night temperature drops to less than 70 degree Celsius below zero.

Venus is roughly of the size of the Earth. Its gravity is about 90% of that of the Earth. It has a dense atmosphere, which has more than 96% carbon dioxide while the rest is nitrogen and sulphuric acid. The sunlight heats its surface, which then radiates this heat as infrared rays. Carbon dioxide, being a greenhouse gas, absorbs most of the infrared rays radiated by the surface of the Venus and radiates it back, heating it to more than 462 degree Celsius and making it the hottest planet in the solar system.

Immediately after its formation, the Earth was very hot. Its surface was molten, and it had almost no atmosphere as most of the light gases escaped its gravity. As Earth cooled, there were severe volcanic eruptions, which released vast amount of hydrogen sulphide, methane, and carbon dioxide. The methane gave rise to a hazy sky. As the Earth's surface cooled and solidified, water started to collect on it, brought to it by icy comets, asteroids, and planetesimals, and oceans were formed. Initially, there was no free oxygen on the Earth except in compounds like water. It is believed that complex reactions taking place near the hot vents inside oceans, made simple living cells which did not need oxygen. And then some organisms called cyanobacteria started making oxygen using carbon dioxide and sunlight, through photosynthesis.

As enough oxygen was produced, it reacted with methane and cleared the skies, and opened the way for sunshine. The oxygen also interacted with iron in the crust to make iron oxide and give rise to red earth which covers vast stretches of Earth today. The oxygen led to the evolution of life forms that could use oxygen to create energy. These forms mostly used the oxygen dissolved in water.

Fig. 3.11: Life May Have Evolved Near Thermal Vents in Deep Oceans [34]

It is believed that about 430 million years ago, life on land appeared. Small plants and invertebrates (animals without backbones) evolved the ability to live on land and use oxygen directly from the atmosphere. During the period of around 416–397 million years ago, plants evolved, as did the first four-footed animals.

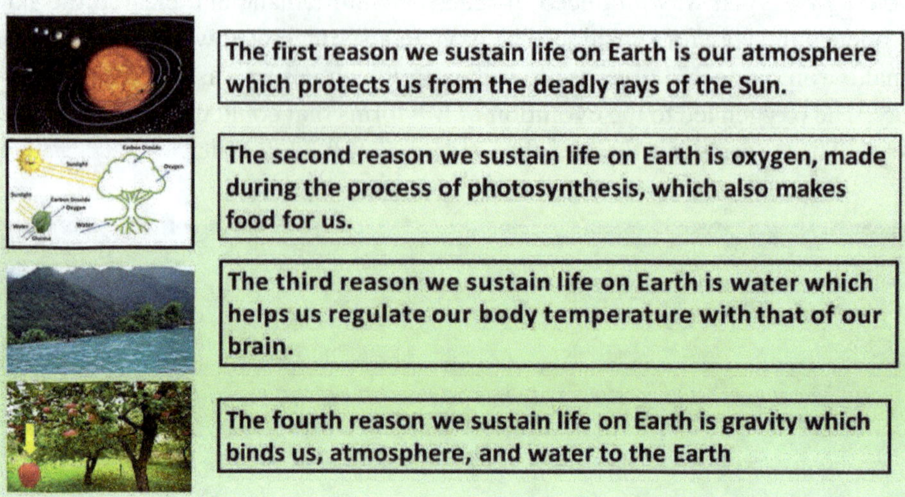

Fig. 3.12: How is Life Sustained on Earth?

As plants occupied land, they produced more and more oxygen and almost the entire Earth was covered with great forests and large swamps of lands. The dead trees and animals dropping into these swamps got converted into coal and oil under pressure, heat, and chemical reactions. These are now fossil fuels. We add though that some scientists have speculated that life on Earth was seeded by comets or meteorites (from Mars, Venus, or outer Space).

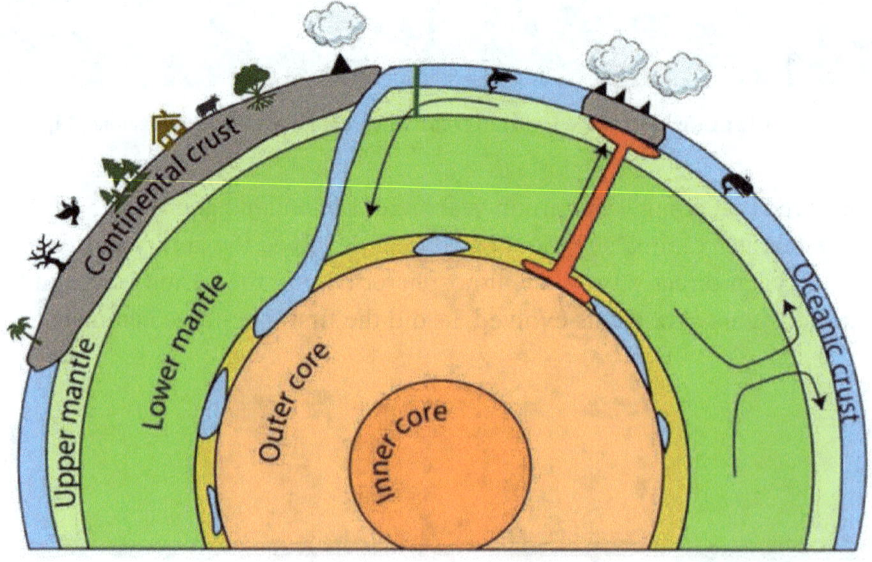

Fig. 3.13: Cross-section of the Earth [35]

Now the Earth has four main layers. The solid inner core of iron has a radius of about 1200 kilometres and a temperature of the order of about 5500 degree Celsius. The normal melting point of iron is about 1540 degree Celsius, but it is solid in the inner core due to tremendous pressure of about 3.4 million times the atmospheric pressure on the surface of the Earth. An outer core surrounds the inner core and has a thickness of about 2300 kilometres. It is made up of liquid iron and nickel and is surrounded by an envelope of the mantle and the crust. Mantle lies between the crust and the outer core and is the thickest layer. It is a hot and viscous mixture of molten rock and is about 2,900 kilometres thick. It has the consistency of caramel. The outermost layer is Earth's crust. It could be up to 30 kilometres deep on an average on land. However, at the bottom of the ocean, the crust is much thinner and may extend to only about 5 kilometres from the sea floor to the top of the mantle.

The atmosphere of Earth has evolved with time. Presently, Earth has an atmosphere consisting of 78% nitrogen, 21% oxygen, 0.9% argon, and about 0.04% carbon dioxide, along with other greenhouse gases. It should be remembered that this small amount of carbon dioxide has played a very important role in keeping Earth's average temperature to a very comfortable 15 degree Celsius. In the absence of carbon dioxide, the temperature of the Earth would have been 18 degree Celsius below zero, and the life as we know would not have evolved. And, equally importantly, the Earth has a vast amount of (liquid) water in its oceans, which none of the planets in the Solar system have.

The Earth has had cycles of being cold and warm since its formation. It is generally believed that there was a time on the Earth around 700 million years ago, when the Earth was so cold that the entire Earth — right up to the equator was covered in ice, as there was no carbon dioxide in the atmosphere. This period is called the period of "Ice-ball Earth" or "Snow-ball Earth". The initial release of carbon dioxide due to volcanic eruptions slowly raised the temperature, utilizing the greenhouse gas effect.

Fig. 3.14: Greenhouse Gases Heating the Earth [36]

One must remember that the molecules of nitrogen (N_2) and oxygen (O_2) consisting of identical atoms do not form an electrical dipole when stretched and have only weakly coupled vibrational levels in the infrared regions. The Earth absorbs the sunlight and radiates it as infrared rays, which are absorbed by carbon dioxide and other greenhouse gases — exciting them to one of their vibrational levels. These are radiated back to the Earth (though some of it escapes into the space), heating it in turn.

The upper layers of atmosphere of the Earth have ozone (O_3). Highly energetic ultraviolet rays from the Sun split the oxygen molecules (O_2) into highly reactive oxygen atoms (O), which combine with oxygen molecules to form ozone. The remaining less energetic ultraviolet rays from the Sun are absorbed by ozone which again splits into an oxygen molecule and an oxygen atom. This protects all plants and animals from the very harmful effects of ultra-violet rays and has played an important role in making life possible.

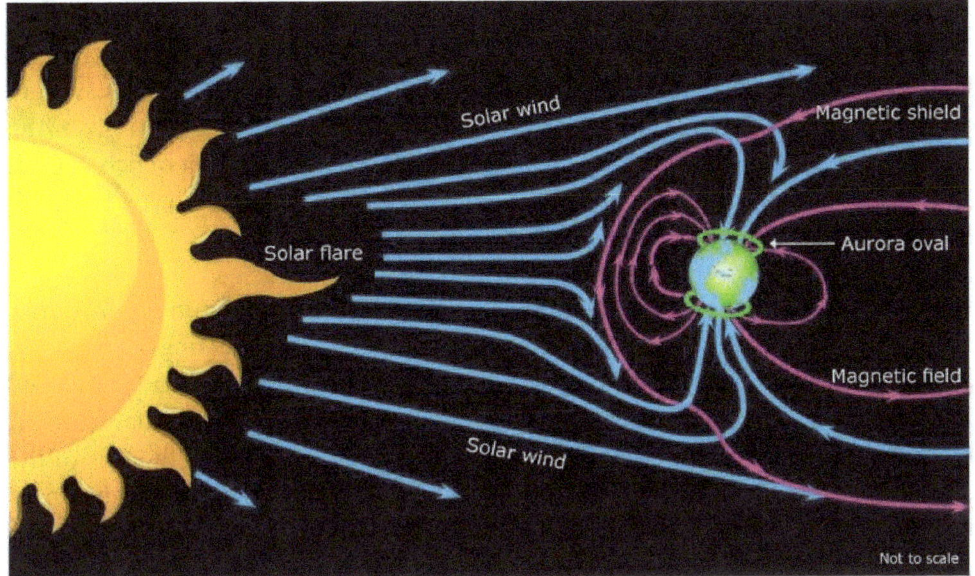

Fig. 3.15: Earth's Magnetic Field Protects it from Solar Flares [37]

The atmosphere of the Earth also protects us from incoming meteorites, most of which break-up due to heat generated by friction before they hit the Earth.

The Earth also has a strong magnetic field of its own. Its magnetic field protects Earth from Solar Flares and the shower of cosmic rays, making life possible. It is produced by rotation of molten iron present in the outer core of the Earth. The necessary heat for the melting of the iron is provided by the radioactive decay of elements Potassium-40, Thorium-232, Uranium-235, and Uranium-238. We add that now it is believed that heavy elements like gold, platinum, and Uranium were produced in collisions of neutron stars, which were formed in supernova explosions. On the other hand, Mercury has only a very weak magnetic field, while Venus and Mars have no magnetic fields at all.

Thus, we are most fortunate to be living on Earth. It is neither too hot, nor too cold. It has water. It has sunshine. It has plenty of carbon, nitrogen, oxygen, and hydrogen. Carbon bonds easily and carbon dioxide — the food for plants, is a gas. Our Earth lies in the so-called Goldilocks zone and several factors listed above have made life possible.

Enter the Moon

It is believed that soon after the formation of the Earth, about 4.3 billion years ago, a large object of the size of Mars, hit the Earth. This object has been given the name Theia. The collision produced intense heat and a huge amount of debris from Theia and the Earth was thrown into space. This debris continued to orbit the Earth and coalesced to form the Moon!

It also did something else, which had a profound effect on the future of the Earth. It set the Earth spinning on its axis. This resulted in a uniform heating of the Earth, and day and night. In the absence of this, the side of the Earth facing the Sun would have been too hot, and the one away from it would have been too cold, making it difficult for life to evolve. It also gave a tilt of about 23 degrees to the axis of the Earth! We know that the Earth has a wonderful variation of the seasons due to its tilted axis, which is important for cultivation of food grains and fruits.

The Moon (along with Sun to a much lesser extent) is also responsible for causing low tides and high tides in the oceans of the Earth — which maximise the interaction between energy (sunlight), water, land and air — leading to conditions suitable for life.

Over the last millions of years, life has spread to occupy every corner of this planet. Now we have forests — teeming with wildlife and colourful birds, we have seas and other waterbodies — teeming with marine life of every size, shape, and description. And we have us, humans — or Homo sapiens. Humans have built cities to live, cultivated land to feed billions, domesticated animals, and birds, and have conquered their physical limitations to conquer Earth, skies, and oceans and have started to explore space.

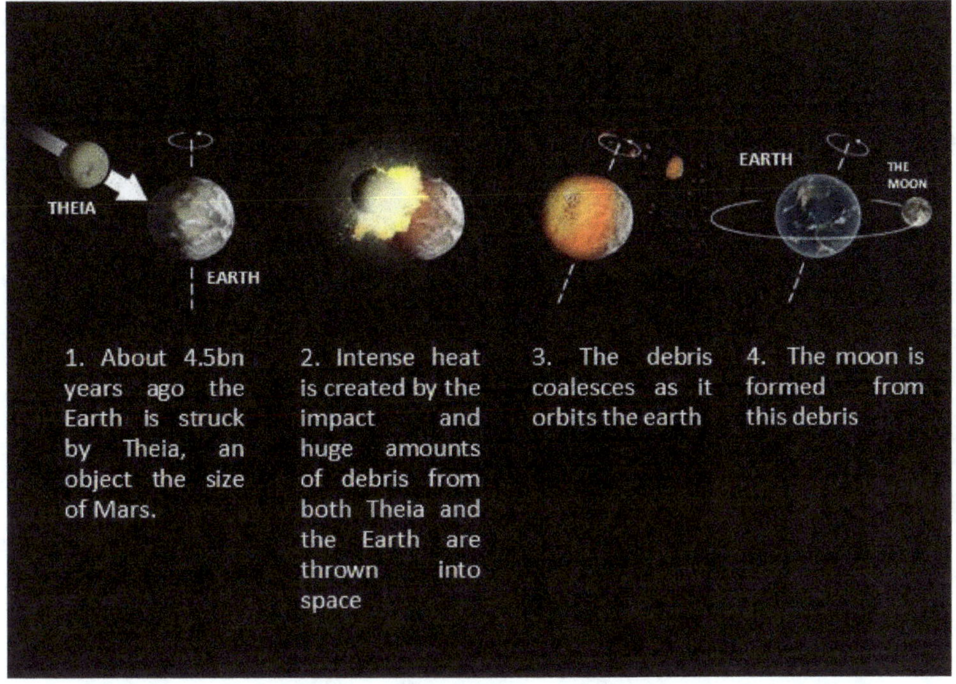

Fig. 3.16: Birth of the Moon [38]

The Wonder that is India

By now it is accepted that till about 140 million years ago, South America, Africa, India, Madagascar, Antarctica, and Australia were parts of a landmass which has been given the name of Gondwana Land. The name Gondawana — Forest of Gonds, the land of Gond tribes of Central Northern India, was given by Eduard Suess after glossopteris fern fossils were found there. This landmass broke apart due to massive movements and breaking up of tectonic plates. Antarctic Beech trees found in South West Australia are also known to have a common origin to similar trees in some of the countries of the Southern continent, e.g., Argentina, Chile, and New Zealand.

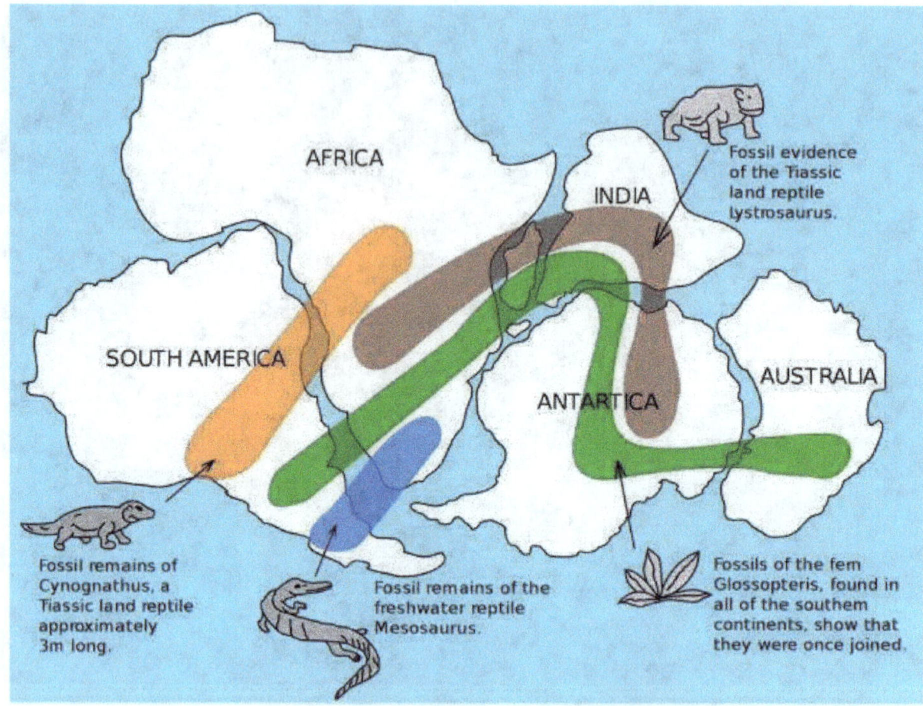

Fig. 3.17: Gondwana Land [39]

Fig. 3.18: Fossils of Glossopteris Flora [40]

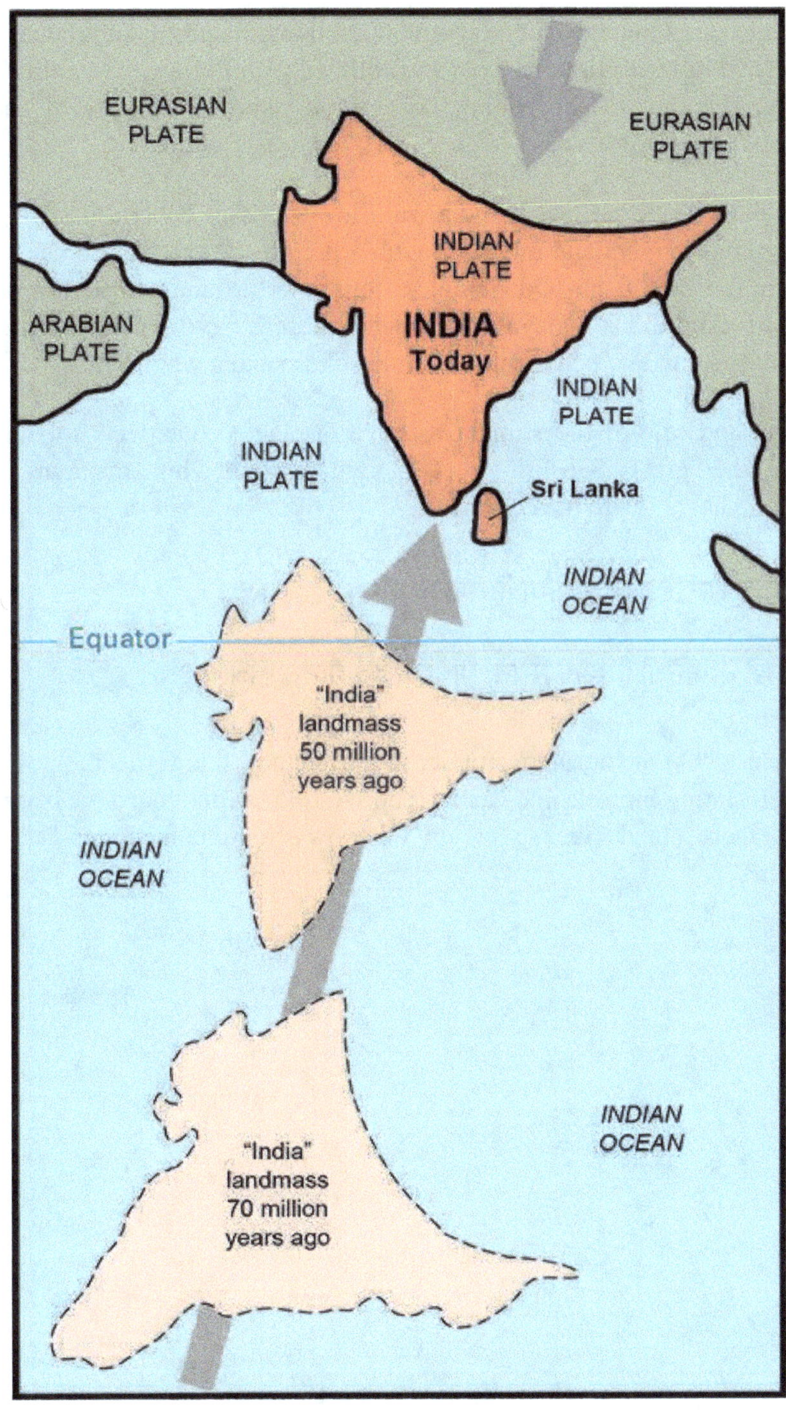

Fig. 3.19: India arrives from Antarctica! [41]

Indian plate completed its drift of about 3000 kilometres from Antarctica, moving at a speed of 5 to 15 centimetres/year and collided with Eurasia. The collision gave rise to the Himalayas. The Himalayas are still rising at the rate of about 2.5 centimetres/year.

The Himalayas have protected India from icy winds of Siberia. It has given it several perennial rivers and monsoon with a precise pattern. These have made a land rich in flora and fauna and vibrant seasons, a fertile land irrigated by rains and many rivers. And the Himalayas, along with the seas have provided it protection against invaders and helped create a unique culture and a wonderful civilization.

The charm and role of the Himalayas have fascinated our poets for millennia. Kalidasa, the great Sanskrit poet started one of his magnum opuses, Kumarasambhavam with the comment:

अस्त्युत्तरस्यां दिशि देवतात्मा हिमालयो नाम नगाधिराजः।

पूर्वापरौ तोयनिधी विगाह्य स्थितः पृथिव्या इव मानदण्डः॥ १-१ ॥

which means, "On the northern frontier of this country that forms the heartland of gods, intercalating himself into eastern and western oceans like a measuring stick of Earth, there stands the sovereign of snowy mountains renowned as The Himalayas".

Chapter 4

The Incredible Transformation of Hunter–Gatherers into Civilization Builders

It is generally believed that the first human ancestors appeared between five and seven million years ago, probably when some apelike creatures in Africa began to walk habitually on two legs. This helped them look up to larger distances for dangers and food and freed their two hands to manipulate objects. They were already flaking crude stone tools. Some of them even spread into Asia and Europe. The early man had to compete with other animals not only for his very survival but also for his food, the source of his energy and strength. His foods had their origin, directly or indirectly, in the photosynthesis from the insolation of the Sun. The discovery of Fire, Agriculture, and Industrialization has been recognized as major milestones in the evolution of the humans on Earth.

The discovery of fire and its control not only widened the range of food the humans could eat but also gave him warmth helping him expand his region of settlement to larger areas. Cooked meals took less energy and time to digest, which left him with more energy to explore wider regions. Fire also helped him to smelt copper, tin, and iron — which helped him to go beyond the stone tools. The metallic weapons helped him to "conquer" animals more easily. This "hunting and gathering" stage in man's evolution could of course sustain only a limited population.

Agriculture was the next paradigm shift in man's evolution which started around 10,000 years ago. Man cleared vast areas of forests for agriculture. Man had already domesticated animals and discovered wheel by then. Now the planet could support a much larger population. Villages, leaders, religions, cultures, *etc.*, all developed during this phase marking the beginning of civilizations as we know them, nearly 6000 years ago.

However, one thing had not changed. Man had access to only that energy which the Sun was providing every day and that was being converted to trees, plants, fodder, and food. This energy was fixed. The variations of weather, like droughts, heavy rains, more severe spells of cold, *etc.*, must have resulted in severe stress both for men and animals. The "poorer" sections of the society must have suffered more. Let us look at the increase in the power brought to man by animals in a little more detail.

A convenient unit of power, often used even at present, is horsepower (HP), which amounts to lifting a weight of 550 pounds to a height of one foot in one second.

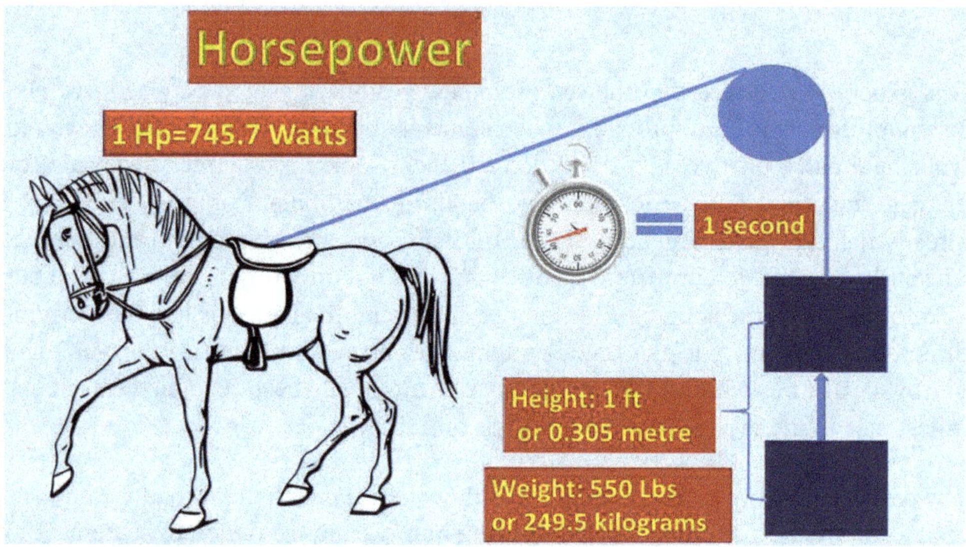

Fig. 4.1: Horsepower [42]

It is equivalent to 745.7 Watts of electric power of today. It is interesting to look at the power of other domesticated animals which became available to man.

Table 4.1: Power of Animals and Man

Animal	Power (Watts)
Horse	746
Draft Horse	645
Ox	430
Mule	320
Donkey	160
Man	67

Thus, using animals, the power available to man could be increased by many folds, with an ox doing the job of more than six people. This coupled with wheels and riding horses, literally gave wings to man, who went on to conquer the entire planet.

Fig. 4.2: Oxen Put to Work in India

It is interesting to recall that in the absence of horses, draft animals and wheels, the Native Americans used travois, sometimes pulled by dogs to travel with their belongings.

Fig. 4.3: Travois used by Native Americans [43]

During this period, all the food and most of the materials of common use were produced by agriculture. Thus, the availability of cotton, wood and a large variety of raw materials was decided by the productivity of land. Even the production of metals was limited by the supply of wood and charcoal, which were used for smelting. As the land and its productivity were limited, the carrying capacity of the Earth was reaching a plateau. It is not surprising that during the agricultural phase, the bulk of the population remained poor compared to the standards to which we are used to now.

Slavery — A painful chapter in human history

Before proceeding, we must take a brief pause and look at a very painful and a very long-lasting episode of human history, which caused enormous sufferings and displacements of mankind and whose scars have not yet healed. Slavery, as a source of manpower or labour, was rampant in almost all civilizations and taking vanquished people as slaves following wars was a common practice. The slaves were mostly used to provide labour, though sexual slavery was also quite common. These are not unknown even now.

The Mongols used an even more brutal form of slavery as they always slaughtered all but artisans of the defeated nations. Slaves taken from neighbouring villages were extensively used by them as human shields during raids on forts in China, when the people inside could see their relatives being goaded to move ahead of the invading Mongol armies.

It is difficult to find accurate estimates of Indians taken as slaves to Central Asia and Persia by invaders, but it could easily be several millions, over hundreds of years.

It is estimated that more than 12.5 million Africans were taken as slaves to Americas, and about 2.5 million may have died during the sea voyage and thrown over-board. They were made to work in sugar plantations and cotton fields earning enormous profits for their owners.

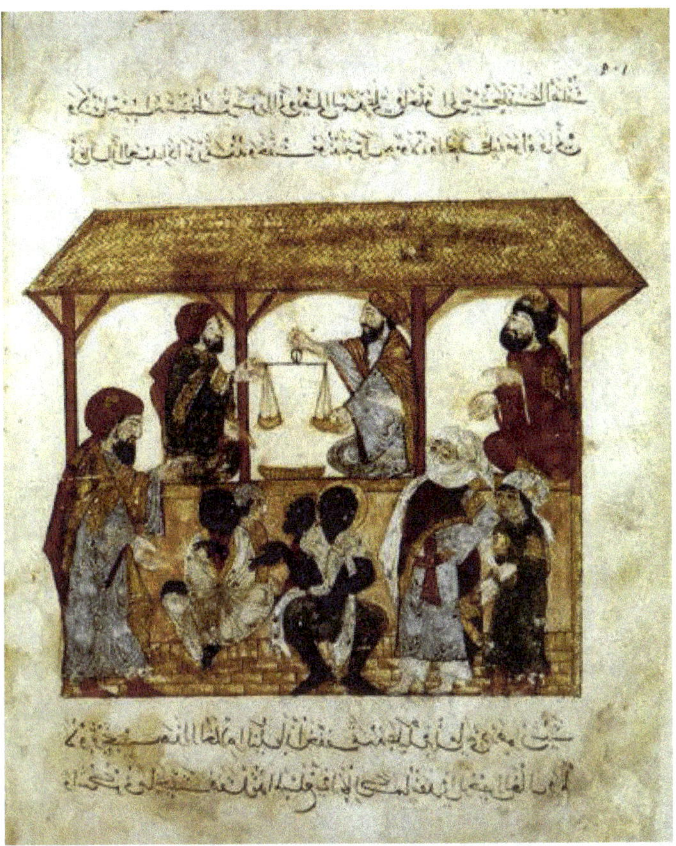

Fig. 4.4: A Slave Market in Yemen [44]

Fig. 4.5: Slaves from Africa being Taken to Americas [45]

It was common for Portuguese and Dutch to raid villages along the Hooghly river for capturing people to be sold as slaves. Slavery was abolished in the British Empire, which included North America, in 1833, in French colonies in 1848, and in Dutch Empire in 1863.

To meet the demands of labour, an indenture system of loan-bondage was introduced which took away more than 2 million Indians to Caribbean, South Africa, Mauritius, and Fiji, *etc*.

Theirs was a very tough life, with more than 16 hours of work, insufficient food, frequent and brutal punishments, and exploitations.

Fig. 4.6: Indentured Indian Labourers in Trinidad [46]

Mechanization of agriculture with a variety of engines, resulting from the Industrial Revolution, brought this brutal form of suffering for a large part of the world population to an end. However, debt bondage may still be affecting a few million people in the World.

While the Industrial Revolution of the eighteenth century was a turning point in the history of human civilization, the evolution has not been smooth.

Collapse of Civilizations

Marcus Garvey, the famous black activist from Jamaica, once remarked, "A people without the knowledge of their past, history, origin, and culture is like a tree without roots." Thus, indeed the study of history of civilizations is one of the most fascinating and rewarding branches of knowledge. We know that Homo sapiens, having originated in Africa, expanded to occupy the entire world. Taking advantage of the prevailing conditions at the time of their migration — like Bering land bridge between Asia and Americas, they crossed over to North America. Similarly, taking advantage of lower sea levels (by 150–200 metres from their present value) and small boats, they populated Indonesia, Malaysia, Australia, and islands in the Pacific. They developed distinct cultures, ways of life, agriculture, food habits, languages and customs, religions, and mythologies. Some of these civilizations rose to great heights and then collapsed. The reasons, as far as we can see now, are quite varied.

Wars

Some civilizations succumbed to wars to control larger areas of land for agriculture, animals, wealth, and even slave force to work for the victors — so that they had access to more energy available to them and could live a life of comfort and luxury. It is sometimes said that wars used to be lost by a population which had become soft by a life of comfort and luxury to "barbarian" invaders.

Mythologies of almost all nations have countless references to wars, some of which could have had their origins in tribal memories of ancient wars. The practice of putting the entire population of the vanquished to sword was followed by many victors throughout the known history. Alexander the Great sacked and destroyed the city of Persepolis after his victory over Darius III. He burned the Great Palace and the surrounding city to ground and destroyed hundreds of years' collection of writings and art along with magnificent palaces and halls. Genghis Khan's army is believed to have killed up to 40 million people from Asia to Europe. This included several tens of millions in China alone. Up to three-fourth of the population of the area covered by modern day Iran was put to death. When his grandson, Hulagu Khan sacked Baghdad, he ordered the slaughter of about one million people. The Grand Library of Baghdad, was burned, destroying countless precious historical documents and books on subjects ranging from medicine to astronomy. It is said that the waters of the Tigris river turned black with the ink from the books. A bridge wide enough for his army to cross the river on horseback was created by dumping the books in the river.

Scholars have estimated that the military campaigns of Timur Lung killed up to 17 million people and laid sizeable parts of Asia, Africa, and Europe to waste. He is believed to have ordered the slaughter of at least 100,000 people in the city of Delhi. His attack lasting 15 days changed the city into a place of bloodshed, ruin, and destruction. He carried away immense spoils of rubies, diamonds, pearls, gold, silver, and slaves. He took away all the artisans, who went on to create wonderful palaces, mausoleums, and places of worship in Samarkand. It is said that, wherever he went, he brought about destruction, massacres, burning, looting and dishonour to women. He destroyed palaces, places of worship and agriculture, leaving behind disease and famine.

The number of people killed in First World War is estimated as 20 million, while the Second World War took the life of 70–85 million people and flattened innumerable cities.

Fig. 4.7: Pompei and Vesuvius [47]

Natural Disasters

Traditions in many nations talk of destruction of entire cities due to natural disasters like earthquakes, volcanic eruptions, and floods. Pompei, along with the neighbouring city of Herculaneum and the surrounding areas, was buried under 4 to 6 metres of volcanic ash and pumice in the eruption of Mount Vesuvius in AD 79.

Fig. 4.8: Banda Aceh, Indonesia after a 9.0 Magnitude Earthquake (2004)

Bhuj (India, 2001), Bam (Iran, 2003), Banda Aceh (Indonesia, 2004), Port-au-Prince (Haiti, 2010), Minamisanriku (Japan, 2011) are some of the cities destroyed by earthquakes in recent years.

Plague or the Black Death, as it was called, killed 75 to 200 million people in Eurasia during the fourteenth century. Several tribes in Americas were wiped out by smallpox — to which they had no resistance when they met Europeans. Spanish Flu infected up to 500 million people across the world during 1918–1920. The estimates of death toll vary from 17 million to 50 million and some as high as 100 million.

Swarms of locusts have appeared at irregular intervals throughout the history, some as large as 500,000 square kilometres (Midwest, USA, 1875), which "blocked out the Sun" and devoured everything, "fields of crops, the leaves on trees, every blade of grass, the wool off sheep, the harnesses off horses, the paint off wagons, the handles off pitchforks, fabrics, clothing…". They are believed to emerge after a good rain following a prolonged drought.

Climate Change and Overexploitation

We cannot control natural disasters. The avoidance of war must wait for the emergence of a more just society and much more enlightened humans.

Let us concentrate on climate change, which is our immediate concern. At the risk of repeating, we note that the continued survival of an organism or a civilization is decided by what we called the "carrying capacity" of the environment, which is decided by the amount of necessary resources available.

Imagine an island covered with vegetation, but which has no predators for rabbits (recall Australia!). If we introduce rabbits there, they will eat as much as they can and reproduce rapidly. Soon, the rabbits would finish most of the vegetation and reach the "carrying capacity" of the island. Now their population would rapidly fall, and they will ultimately face extinction due to starvation and disease. A recent example, which supports this argument, is overgrazing of steppes in Mongolia by goats, which is leading to expansion of the Gobi Desert.

Nature has, over millennia, evolved a scheme of "checks and balances" to keep things under control. Humans are also governed by the laws of carrying capacity. It has been argued that when human societies draw more natural resources from their environment, than what is organically replenished every year, it amounts to accumulating an "ecological debt". This debt accumulates, gets "compounded", and leads to the end of the society.

The unbridled growth of a society exulting in consumerism has reached a stage where the end of our society looms large upon us, due to the ecological over-exploitation of the resources of the Earth. We shall come back to this point again, towards the end of the book. We note that most cities and regions are not solely dependent on one geographical area to meet most of their needs.

Pericles, at the end of the first year of the Peloponnesian War (431–404 BC) at the funeral for the war dead remarked that, "Because of the greatness of our city, the fruits of the whole Earth flow in upon us; so that we enjoy the goods of other countries as freely as our own." Thus, for example, U.K. gets bananas from Latin America, apples from France and South Africa, mandarins from Spain, pineapples from Costa Rica, tomatoes and onions from The Netherlands, cauliflower and celery from Spain and potato from France. This globalization of the food system is dependent on the availability of cheap fossil fuels. However, food is just one example. The world civilization, science and technology, commerce, and industry today is driven by globalization, which started thousands of years ago, through trade, exploration, religion, and desire to conquer new lands. We shall have occasion to come back to this. It is instructive to look to the past to learn from societies that drove themselves out of existence by hitting their environment too hard.

The Mayans

The Mayan civilization was a highly developed and sophisticated society in Central America that peaked, in terms of both population and power, in the 8th and 9th centuries. The exact mechanism that caused the group's downfall is still not exactly clear, but most theories cast the blame on climate change.

Fig. 4.9: The Nunnery Quadrangle in Uxmal [48]

Their empire in Mesoamerica, covered the Yucatan Peninsula and modern-day Guatemala, Belize, parts of Mexico, western Honduras, and El Salvador. It lasted for about 3,000 years. They developed writing, art, architecture, calendar, mathematics, and astronomical system. They built great cities during 750–500 BC, which had wonderful large temples and architecture. Agriculture and religion were quite developed with sacrifices — including human sacrifices, as a regular ritual to appease gods and to make the land fertile. It is believed that the overpopulation put a large strain on the resources and led to conflicts with other nations. Large tracks of the rainforest were cleared for agriculture to meet the growing demand. This drove animals, on which Mayans depended for food, to extinction. Finally, an extensive period of drought led to crop failure and drying up of water supplies, leading to the collapse of the civilization.

The Easter Island

Fig. 4.10: Moai Statues in Easter Islands [49]

Easter Island (area about 160 square kilometres), so named as it was discovered by Europeans on Easter in 1722, is situated in the Pacific Ocean between South America and Australia. It is believed to have been peopled by Polynesians around 1200 AD. The Rapa Nui people, as the original natives were called, had chiefs, priests, and workers who fished, farmed, and made the moai statues for which the island is famous. The island had large trees — some as tall as 50 meters — which were cut down to make boats for fishing and to move the statues, leaving no place for birds to nest and nothing for animals to eat. This led to extinction of birds and animals and a crash in the human population from an estimated peak of 15,000 to just about 2,000 when the Europeans arrived. Exposure to smallpox and kidnapping for slavery reduced their number to just about 100.

It is generally accepted that the deforestation had a devastating effect on the people living on the island and was the primary cause of their decimation.

The Sumerians

Fig. 4.11: Sumerian Civilization [50]

The civilization of Sumer or Sumerians was among the earliest civilizations of the world, whose effect is seen till today. It started around 4000 BC and ended by 2400 BC, when the entire area of Mesopotamia was ruled by the Akkadians. It occupied the region of southern Mesopotamia (modern day southern Iraq), between the Tigris and Euphrates rivers. About 200 years later they re-emerged, briefly. The land was very fertile, and they developed farming techniques, like monocropping, large-scale irrigation and use of specialized labour, which are still in use. The abundance of food production accommodated the emergence of a stratified social structure, in which certain classes could indulge in learning of arts, sciences, mathematics, and technology.

Sumerians invented the abacus which helped them add and subtract numbers quickly and developed early arithmetic, geometry, and algebra. Their discoveries in geometry included calculation of area of a triangle and volume of a cube, and an understanding of the Pythagoras theorem. They discovered writing and left behind a vast library of clay tablets — covered in cuneiform script, which have been deciphered and give details of their daily life, accounting, personal and official letters, flow of goods and agricultural produce, and even a large literary work of extreme beauty and imagination, "The Epic of Gilgamesh". They built elaborate structures of temples, palaces, and administrative buildings with terra cotta ornamentation, bronze accents, complicated mosaics, imposing brick columns and sophisticated mural paintings. Several steles have been found which give details of wars, genealogies of kings, and eulogies.

They had commercial contacts with several civilizations like Akkadians in modern Turkey and the Indus Valley Civilization in India. Their discovery of arithmetic contributed to the invention of calendar and clock. They used several number systems including the sexagesimal system (base 60), which later became the standard number system of the Sumerian civilization. The Sumerian's clock had 60 seconds, 60 minutes, and 12 hours; and the calendar had 12 months. We continue to use this system even today to measure time, angles, and geographic coordinates. The Sumerians were divided into several city states, which were constantly at war with each other. This led to the development of military technology, city-walls, the art of siege and even phalanx. It is likely that they invented chariots and used them in battles. One of the greatest advances made by Sumerians was in the field of hydraulic engineering. They created a system of ponds to control flooding, and were also the inventors of irrigation, using regularly maintained canals.

The entire enterprise mentioned above was possible because of the excess food produced by the use of irrigation, which brought larger and larger areas under cultivation. However, the arid climate meant that the irrigation water left behind salt in their soil, as it evaporated. The rise in the salinity each season was minute, but after hundreds of years, they were left with dead fields. Without agriculture to support their population, the Sumerian people melted away from the area and faded into history.

The Indus Valley Civilization

The Indus Valley or Harappan or Sindhu-Saraswati Civilization as it is popularly called, was one of the greatest societies in the ancient world. It was contemporary with Ancient Egypt and Mesopotamia — with which it had commercial exchanges. At its peak, its population is thought to have made up 10% of the world population. It lasted from 3300 BC (or earlier) to 1300 BC and extended from Baluchistan in the west to western Uttar Pradesh in the east, and north-eastern Afghanistan in the north to Gujarat in the south. Sites have been found in Gujarat, Haryana, Punjab, Rajasthan, Uttar Pradesh, and Jammu and Kashmir in India and Sindh, Punjab, and Baluchistan in Pakistan. Several coastal sites like Lothal and Dholavira have been found. The civilization had social hierarchies, writing system, large, planned cities, and long-distance trade.

Fig. 4.12: The Great Bath of Mohenjo-Daro [51]

The cities had the world's first known urban sanitation system. They were among the first people in the world to dig wells and step wells. They built impressive dockyards, granaries, warehouses, brick platforms, baths, and citadels with thick protective walls. They made beautiful beads, sculptures, seals, bronze vessels, pottery, gold jewellery, figurines in terracotta, bronze, and steatite, as well as toys and artifacts for games, including a cubical dice with one to six holes on the sides. They had standard units for length and weight — the later were still in use even in the time of Kautilya, and used copper, bronze, lead, and tin. Unfortunately, their script has not yet been deciphered and the cause of the decline of their civilization is hotly debated. One of the early suggestions for the destruction of the civilization was invasion of war, like nomadic tribes from Central Asia. However, evidence has been accumulating for some time that climate change was possibly the reason for their decay. Some historians suggest that it is quite likely that the now dried-up Saraswati river, along which perhaps many of these cities in the south were situated, either changed course or dried up following a massive earthquake.

It is now generally believed that this wonderful civilization sprung up in a perfect climactic window — which however closed and brought about its eventual downfall. The Indus Valley had vast areas covered by floodwaters from the rivers during monsoons, which left extremely fertile soil in their wake. For over 2000 years, the society was supported by manageable monsoons and the rich soils they brought. They cleared more and more of forests for agriculture, as the population increased. The monsoons slowly shifted to East, perhaps due to extensive deforestation. As the rains continued to lessen, these monsoon floods became unreliable. This eliminated the surpluses and made it impossible for the cities to be supported. The Indus Valley people, thus may have shifted eastward.

Dorian Fuller has opined that, "We can envision that this eastern shift involved a change to more localized forms of economy: smaller communities supported by local rain-fed farming and dwindling streams. This may have produced smaller surpluses, and would not have supported large cities, but would have been reliable…thus cities collapsed, but smaller agricultural communities were sustainable and flourished. Many of the urban arts, such as writing, faded away, but agriculture continued and actually diversified."

The Industrial Revolution

The biggest paradigm shift that took place in the history of human civilizations was the Industrial Revolution that took place in Europe and in United States of America in the period from about 1760 to sometime between 1820 and 1840 with new manufacturing processes when man learned to use machines for production, new chemical manufacturing and the use of steam power. The Industrial

Revolution also led to an unprecedented rise in the rate of population growth and standard of living in the Western world. The discovery of internal and external combustion engines, use of coal and oil to run the engines, discovery of electricity, all added to keep the revolution alive and push the world economy forward, though not equitably. Many resource rich countries found themselves exploited without an active participation in the economic growth.

Even though coal (and crude oil) were known for quite some time, extensive use of coal to provide heat started in good measure only around twelfth century. In fact, it (mostly peat) was originally used by poor people as the forests diminished and they had difficulty in getting firewood.

Fig. 4.13: One of the early Steam Engines called John Bull (c. 1893) [52]

There is evidence that steam was already in use in Greece, China, and Turkey, *etc.* for a long time for doing some simple tasks. One of the first large scale uses of steam engines was to quickly pump water out of coal mines. Extensive use of steam engines started in eighteenth century which started the Industrial Revolution in Britain. The operating principle of steam engine is simple, and was soon adopted for a variety of tasks: railways, steam ships, and textile mills, *etc.*

Coal powered engines unleashed an enormous amount of energy for man — which he used to increase the production of goods by orders of magnitude.

Production during the agricultural phase was limited by the supply of energy captured by annual amount of energy captured by the plants using photosynthesis. Interestingly, plant photosynthesis has continued to be the dominant source of energy in the era of the Industrial Revolution. However, it was (and continues to be) energy drawn from many millions of years of plant photosynthesis in the form of coal. Thus, the energy accumulated over millions of years by the plants became available to man. The "honours" were later extended to oil, which was the accumulated product of marine animals which died over millions of years. The use of coal ended the laborious poverty of early times. The rise in the coal consumption in England and Wales during this period is a testimony to this.

It would have been impossible to produce the vast quantities of iron and steel necessary for engines, railway lines, and rolling stock, so essential for the success of the Industrial Revolution, without the use of coal. The role of colonies which Britain had acquired by then as a source of raw material and as markets for finished goods is only too well known.

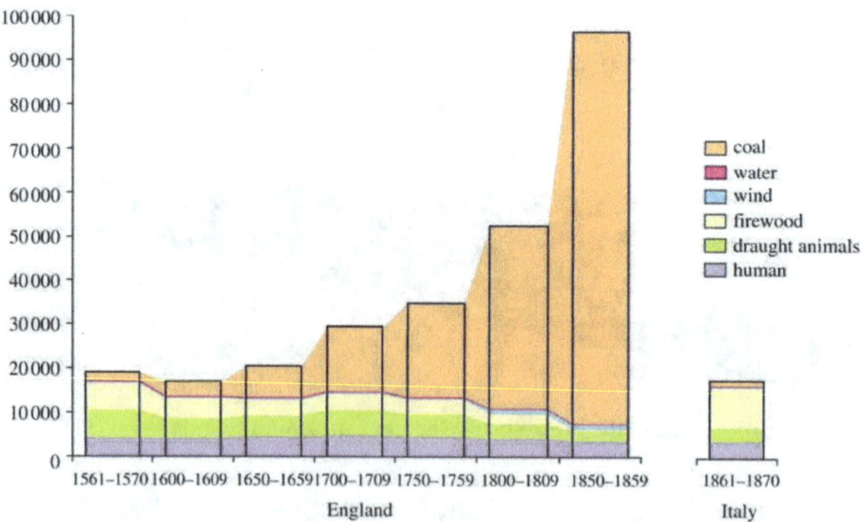

Fig. 4.14: Annual Energy Consumption per head (in Megajoules) in England and Wales [53]

Once internal combustion engines were developed by Carl Benz in 1885 and affordable mass-produced automobiles were introduced by Henry Ford in 1908, oil was used extensively. Oil was enlisted to provide additional source of energy and additional varieties of engines, including automobiles (as mentioned) and aircrafts — which further accelerated the rate of growth. Natural gas soon joined the "party" and has now become extremely popular for production of electricity and for use as fuel. Coal, oil, and gas are all fossil fuels. Now the largest locomotives running on diesel or electricity have powers ranging from a few thousand kilowatts to several tens of thousands of kilowatts. One of the largest ships of the world has a power of about 109,000 Horsepower or more than 80,000 kilowatts.

Thermal power plants of several hundred Megawatts each, hydroelectric plants of more than 10,000 Megawatts, solar plants of very large capacities and nuclear reactors of more than 1000 Megawatts each are available across the world. These numbers are given, only to indicate the long journey that man has made since "the devaluation of his muscle power", when he had access to just seventy Watts of power of his own muscles.

The Science and Technology Revolution of the Twentieth Century

Twentieth century was yet another turning point in the history of human civilization. Not only the early decades of the twentieth century witnessed several path-breaking scientific discoveries that altered our understanding of the material universe but also several new technologies arising out of these scientific discoveries. Nuclear technology, semiconductor technology, genomics and lasers are some of the path breaking technological breakthroughs of this period. Every new technology resulted in several new products and services to the common man. It would not be an exaggeration to say that today there is no aspect of human life untouched by new technologies in some form or another. The computer and communication technologies have made information available on demand anywhere, anytime to anyone. Together with the breakthroughs on transport technologies, the world has indeed shrunk into a global village.

Energy is the backbone of all technologies and it is not surprising that energy needs of the world continues to rise. While world has evolved into a global village, inequalities have also grown, leading to developed, developing and under-developed, the technology rich, the technology poor countries, and the energy rich and the energy poor segments of population.

It is interesting to put on record that, Robert Lucas, Jr., who received Nobel Prize (1995) in economics, had noted that, "For the first time in history, the living standards of ordinary people underwent an unprecedented and sustained growth, during the Industrial Revolution."

We still use fossil fuels to meet a large part of our energy needs, as we can see from the graph below.

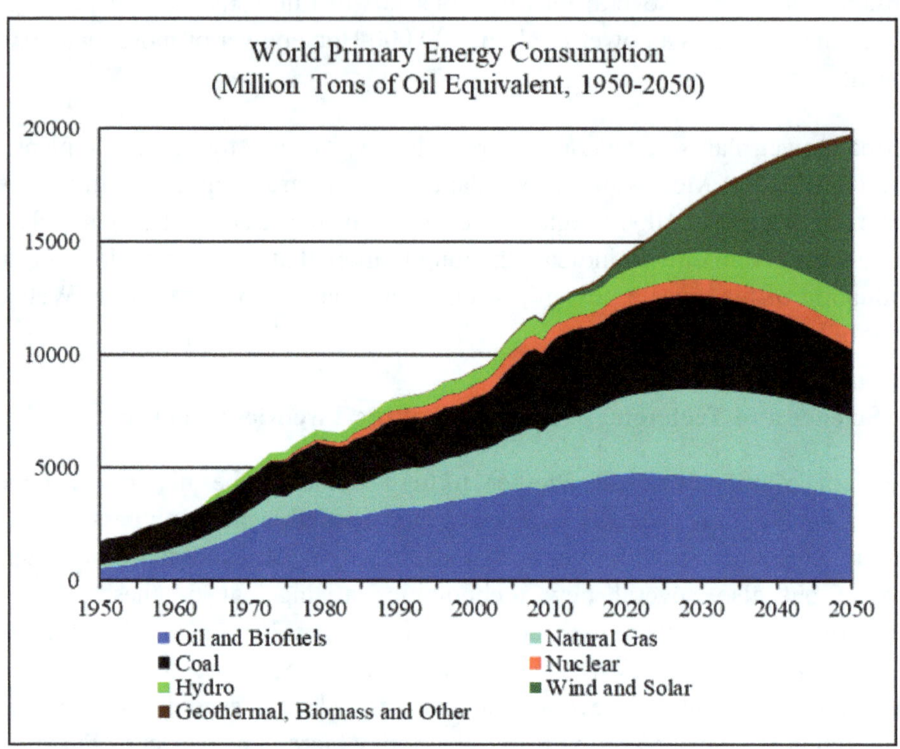

Fig. 4.15: World Primary Energy Consumption, 1950–2050 [54]

This, however, has had an unwanted and deleterious effect. We have seen that, from the time when life appeared on Earth, it has endured several disasters like Ice Ages, mega earthquakes, mega volcanic eruptions, splitting, drifting away and collision of continents, and giant asteroid impacts. It has recovered from each one of these over a course of time. All the greenhouse gases, responsible for keeping the moderate temperatures of the Earth were kept in check by the ecosystem, and the carbon dioxide for example, had never ever reached a value larger than 300 ppm, during the previous 800,000 thousand years, which we can infer from air-bubbles trapped in ancient Antarctic Ice (see later, for details).

This level of carbon dioxide is continuously on the rise since the beginning of Industrial Revolution. And it is leading to global warming and climate change. It is highly destructive for the ecosystem of the Earth. If steps are not taken immediately to stop and reverse this growth of greenhouse gases, our very existence could be in doubt.

In what follows, we shall examine the human development, the world energy requirement in years to come and the global warming that has been caused by extensive deployment of fossil fuels. We shall also discuss various sources of energy available to mankind in pages to come along with their advantages and disadvantage. We shall see that renewable sources of energy, e.g., solar and wind and nuclear energy can mitigate the global warming and the climate change — swiftly and effectively — without compromising with the quality of life to which man has become accustomed and to provide for his expanding future needs as well.

Chapter 5

Global Warming is for Real

We have seen that food, water, and oxygen in the air are the basic needs of life on Earth. Humans learnt to supplement their own muscle power with external help like animal power, machines, *etc.* Engines and electricity are two major breakthroughs — steam and internal combustion engines generated power while electricity ensured power distribution. Hydrocarbons and the biomass were the fuels of choice, as these were easily available and had a large density of energy. Power on demand became a reality in the recent centuries. An unavoidable consequence of sourcing power from fossil fuels is the release of carbon dioxide into the atmosphere.

The idyllic life on Earth, at least where it is not already destroyed by man-made strife, abject poverty, and suffering, is now facing a serious challenge due to global warming, caused by the activities of its most recent inhabitants — the Homo sapiens, especially during the last two hundred years. A vast amount of credible evidence has accumulated over the years that the average temperature of the Earth has been steadily rising. Several facts point to this.

The ice cover in Antarctica and Arctic has been reducing at an alarming rate. This has caused a drastic reduction in the population of penguins and polar bears, as they lose their habitat and feeding grounds.

The glaciers all over the world are receding rapidly and many have vanished. Thus, for example, the White Chuk Glacier in Washington retreated by about 2 kilometres in just 33 years and is on the verge of death now.

Nearer home, we now have access to satellite photographs of glaciers of The Himalayas from 1960s and we can compare these to photographs taken in 2000 to see the change.

Fig. 5.1: The Melting of Ice at Antarctica [55]

Fig. 5.2: The Melting of Ice in the Arctic [56]

Fig. 5.3: White Chuck Glacier in 1973 [57]

Fig. 5.4: White Chuck Glacier in 2006 [58]

The ice cover of Greenland is melting at an alarming rate. These are leading to a rise in the level of the oceans. The melting of the ice of the Arctic will not raise the sea levels, as it floats over sea water. However, we should remember that, if all the ice of Greenland were to melt, the sea-levels would rise by 20 metres. And, if all the ice in the Antarctica were to melt, the sea level would rise by 60 meters! This would drown the coastal areas across the world, along with many island nations.

Fig. 5.5: A glacier in Tibetan Himalaya: 1921 (upper panel) and 2008 (lower panel) [https://sandrp.in/2014/08/20/climate-change-himalayan-glaciers-a-news-round-up/]

A study of satellite data has revealed that the mass of ice sheets of Greenland and Antarctic ice sheets has decreased over the years. Data from NASA's Gravity Recovery and Climate Experiment show that Greenland lost an average of 286 billion tons of ice per year between 1993 and 2016, while Antarctica lost about 127 billion tons of ice per year during the same time. The melting of ice reduces the amount of sunlight which is reflected into space and leads to a further rise of the temperature.

The rate of Antarctica ice mass loss has tripled in the last decade. The melting of these ice sheets formed over hundreds of thousands of years has one other serious consequence. There is a vast amount of methane hydrate — formed by ice and methane under the ice cap in Antarctica — the melting of the ice will release this methane into the air, leading to a further concentration of greenhouse gases. There are also fears that viruses lying dormant for millions of years in this ice may also be released, leading to massive infection of marine life.

The Himalayas have acted as towers of fresh waters for the Indian subcontinent and parts of China and neighbouring countries since time immemorial. The drying up of glaciers in The Himalayas will affect the life of a vast population in these countries — as several perennial rivers like the Ganges, the Brahmaputra, and the Indus would be affected. The glaciers in the rest of the world, e.g., in the Alps, the Andes, the Rockies, Alaska and Africa, are all showing signs of retreat to various degrees. Observations by satellites reveal that the spring snow cover of the northern hemisphere of the Earth is decreasing every year and the snow is melting earlier.

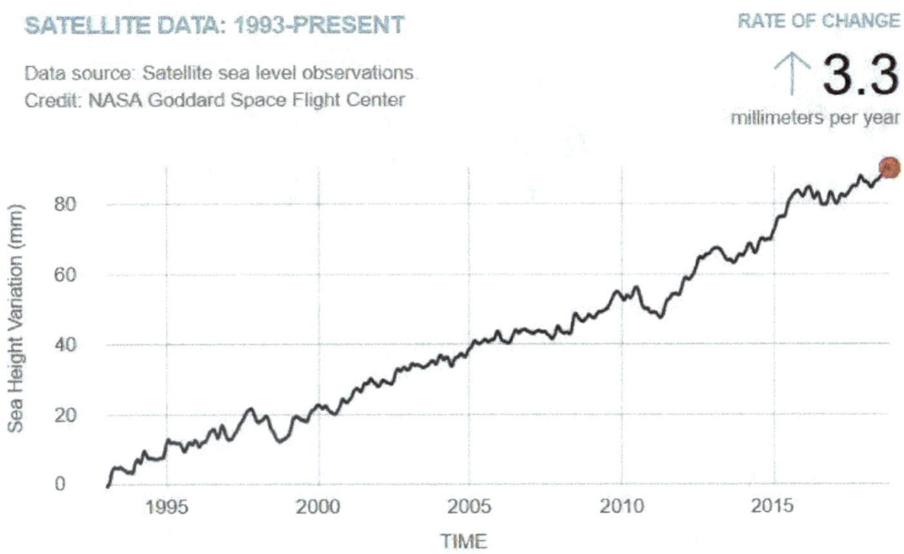

Fig. 5.6: The Rise in Sea Level [59]

We have evidence that the average sea level has been rising consistently since about 1800, a date which coincides with the beginning of the Industrial Revolution.

It is estimated that the sea level has been rising at the rate of 1.2 to 1.7 millimetre per year leading to an increase of about 13 to 20 centimetres from 1900 to the year 2000. This rate of increase is more than 3 millimetres per year now.

A part of this rise is due to rising temperatures of the sea water, which causes it to expand.

Fig. 5.7: Severe Flash Floods in Europe

The rise in sea temperature has several serious consequences. It leads to more severe and frequent cyclones, storm surges, and coastal erosion. It is responsible for excessive rainfall in some areas and severe and repeated droughts in other places. It is possible that it can seriously affect and alter some streams in the oceans in a complex manner.

Fig. 5.8: Explosion in the Population of Jelly Fish [60]

Global warming and associated heating of the oceans has led to a considerable increase in the population of jelly fish and they are now found much further north and south of the equator — in areas which were traditionally considered to be too cold for them. While other fish die in polluted waters, which is often deficient in oxygen, the jelly fish seem to thrive there. It also happens partly because their competitors for food die in such waters.

Nearer home, the rise in the temperature of Arabian Sea is already causing a severe reduction in the catch of oil sardines and mackerels near the coast of Kerala, as the fish have either moved to lower depths or to colder regions up the west coast, near Goa and further north. Further, four cyclones, which are rather rare on the western coast, hit it during 2019. Fortunately, India has a very robust and accurate forecasting system and preparedness for cyclones in place now, which has essentially eliminated loss of life.

We have also seen occurrences of extreme weathers — excessively hot or excessively cold — with too much snow on the rise across the world. The rise in temperature leads to aridity and increasing incidences of forest fires.

California has been experiencing repeated forest fires, which burned down about 260,000 acres of forest in 2019 and caused a loss of about 160 million dollars.

Fig. 5.9: Forest Fires in Australia [61]

The forest fires of Australia in 2019 are estimated to have burnt down more than 25 million acres of forest and killed about one billion wild animals. Many of the animals and birds which perished are unique to Australia. The regeneration of their habitat will take several tens of years. The loss of insect life is much larger, which is quite worrisome for the ecosystem. These also released more than 400 million tons of carbon dioxide into the atmosphere. Fortunately, the loss of human life has been limited to 29, though many people lost their houses and property.

This disaster has brought into focus a very unfortunate asymmetry in the devastating effect of climate change between humans and other living species. It tells us that other living species are much more vulnerable than humans to climate change and will remain so in future when more and more devastations of diverse natures and intensities strike us. It is also evident that ultimately, there will be far less chance of escape for humans as the habitats and space for refuse shrink.

Fig. 5.10: A Satellite Picture of Fires raging across Australia [62]

The shifting of snowline to higher altitudes has had disastrous consequences for apple cultivation in Himachal Pradesh. Insufficient number of cold days, a shortened spring and a longer summer have affected apple production and quality of apples, the world over. Rising temperatures are likely to seriously reduce cacao and coffee yields.

The global warming has had some initial beneficial effects on banana yield though, as it increased where irrigation facilities were available. However, it is suggested that continued rise in the global warming will reduce this yield substantially and may require drastic technological intervention for saving the banana crops.

Fig. 5.11: Recurring Prolonged Droughts in India [63]

The rising temperature is likely to lead to decrease of production of food grains in countries at lower latitudes seriously. It may also lead to severe pest problem as the pests multiply more (they can have 3 life cycles, instead of normal two) as warmer conditions get prolonged. This would require increased use of pesticides, which will have serious consequences for both the pollinating insects as well the human health. There are indications that menace of locusts may increase too. In fact, the nations in and around Horn of Africa and as far as Pakistan and Rajasthan in India experienced invasion by locusts in 2019. This menace continued in 2020 as well, as it was already predicted by researchers.

As the plants may grow with more sugar because as more carbon dioxide is available, there are fears that it may lead to deficiency of nutrients in the food grains and fruits produced. Paradoxically, it may lead to better yield of food grains in countries at higher altitudes. Thus, global warming affects the poorer countries, who contribute the least to it, most harshly, and who are ill-equipped to deal with it.

The rising temperature of the oceans allows them to absorb more carbon dioxide. It is estimated that oceans have absorbed about 525 billion tons of carbon dioxide produced by humans since 1850 or industrialization. They continue to absorb up to 25 million tons of carbon dioxide per day, which is about a quarter of the carbon dioxide emitted everyday by us. Had it not been so, the effect of global warming would have been a lot more severe and it would have hit us much harder and much earlier. It has increased the acidity of the oceans by about 30% during the above period, strongly affecting the population of a large variety of fish.

Carbon dioxide reacts with seawater to form carbonic acid, which releases hydrogen ions (H^+). These hydrogen ions combine with carbonate ions (CO_3^{-2}) to form bicarbonates (HCO_3^-), which reduces the amount of carbonate ions in the water. It is estimated that the ocean carbonate concentration has decreased by about 10% since industrialization. The scarcity of carbonate ions makes it difficult for corals, oysters, clams, mollusks, and other shelled organisms to make shells. This then affects the entire food chain in the oceans, as up to 4000 species are dependent on corals.

We have seen that the temperature over the land has been rising, the sea surface temperature has been rising, the sea levels have been rising, the polar ice cover has been decreasing, the snow cover is decreasing, and the glaciers are retreating. We also realize that this is also necessarily leading to rise of humidity, rise of temperature over oceans and rise in the temperature of the air near the surface of the Earth.

Additional indications of global warming have come from looking at rings of trees which grow more when temperatures are higher. Butterfly species are rather specific to a narrow zone of temperature in European countries. It has been found that they are shifting their habitats by as much as 250 kilometres to the north. Similar shifting of habitats for birds has also been reported. As the winters are getting shorter, flowers specific to spring have been reported to flower up to two weeks earlier.

All these taken together provide enough clear indications of global warming.

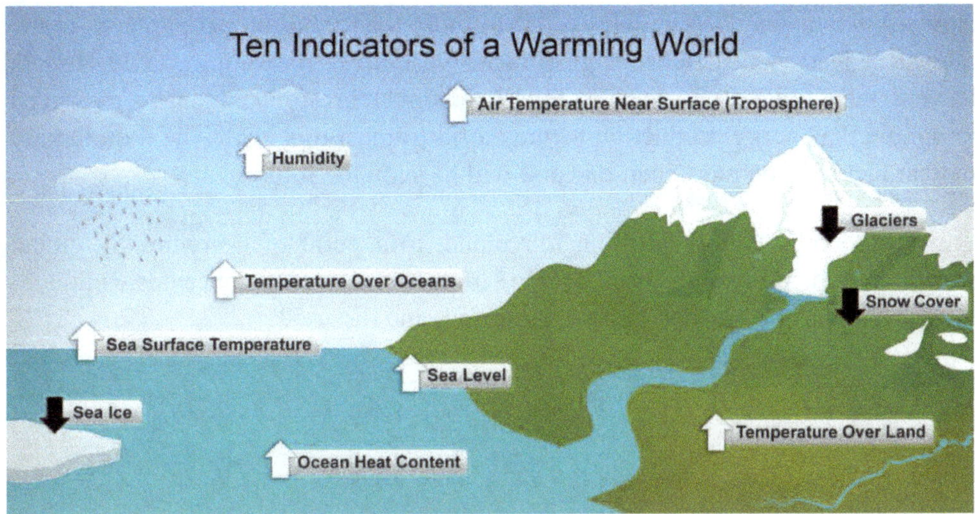

Fig. 5.12: A Summary of Signs of Global Warming, NOAA [64]

Let us be very clear about one thing; the crust of the Earth up to a depth of only a metre or so is used for agriculture, though some trees do draw nutrients and water from up to tens of metres. Till men started making high rise buildings, they only used the atmosphere up to a few metres from the surface and caught fish from not too deep waters.

Now mines can be several hundreds of metres deep, the houses can be hundreds of storeys high, the fish from deep oceans can be trapped and entire atmosphere is open to air transport. This change of widening of layer of the Earth below and above the surface for exploitation by man — that latest arrival on the Earth not too far in the past has opened vast opportunities for our comfortable and healthy living and easy movement. The rates of scientific discoveries are continuously accelerating. But it has come at a cost — we are using energy at a level which was never done in the past. And most of the energy is coming from fossil fuels which were generated by the action of natural forces over several millions of years. And it has come at a severe price.

There is no way that we can overnight go back to days when the Earth had perhaps a very small number of humans, who lived in caves, hunted animals, and gathered food, and most tragically died very young. It has been estimated that the population of Earth in 10,000 BC was perhaps of the order of 2 million.

How are we to provide enough energy to more than 7 billion people now, at the rate of at least 4000 kWh per person per year, to have a comfortable Human Development Index (0.8, see later) for all of them? It is estimated that we may need to double the energy production of the world to meet this goal. The requirement may in fact be even more than this and will be so in the future.

Can we do this without hurtling our planet to a point of no-return by global warming and protecting our people from large scale upheavals, climate-displacements, and ensuing hunger, suffering, and disease?

Chapter 6

Global Warming and Greenhouse Gases

While travelling in North America in 1837 and later, Louis Agassiz noticed rock striations, moraines, and erratic rocks — the kind he had seen in The Alps and proposed the Theory of Ice Age Glaciation. Rock striations are deep scratch marks left by other rocks when these were carried by glaciers over large distances to become erratic rocks. Moraines are debris carried by glaciers to long distances.

Fig. 6.1: Rock Striation, Mount Rainier National Park, Washington, USA [65]

Now we know that the Earth has had multiple Ice Ages. The last one lasted till about 12,000 years ago. During this period, enormous ice masses covered Canada, northern USA, northern Europe, and northern Asia. We also know that humans developed significantly during the last Ice Age. With the end of the Ice Age, the woolly mammoth went extinct.

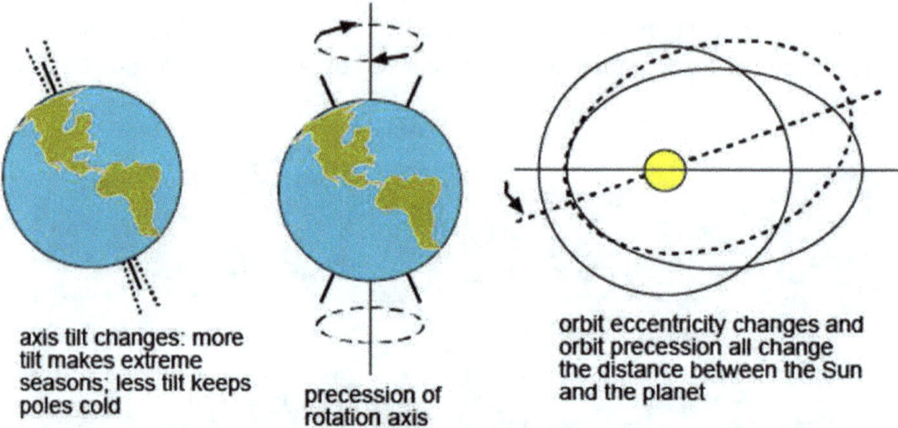

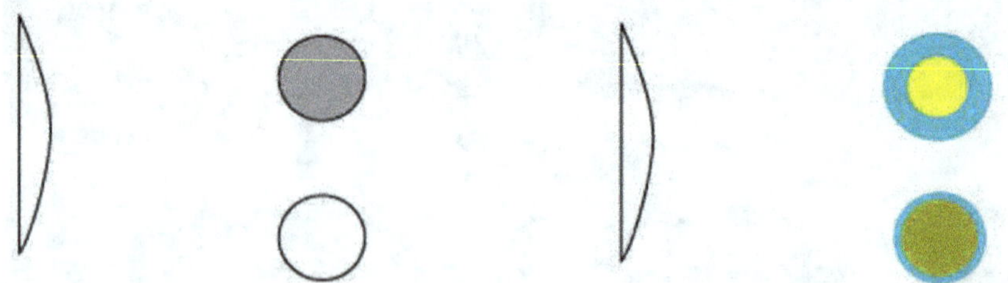

Fig. 6.2: Various Causes of Heating or Cooling of The Earth [66]

Milutin Milankovitch in 1920s had already provided mathematical evidence for such a cyclic behaviour. His calculations, which accounted for the pull of the major planets like Jupiter and Saturn on the orbit of the Earth and some additional effects due to tides, *etc.*, suggested periodic changes in the ellipticity of the orbit of the Earth around the Sun, tilt of its axis, and precession of its rotation axis. The net period for these changes, which increase or decrease the heating of the Earth by the Sun is about 100,000 years.

Fig. 6.3: Ancient Air Trapped in Ice [67]

The other effects affecting the temperature of the Earth could be (periodic) variations in the brightness of the Sun and reflectivity of the Earth's surface — e.g., by its ice cover. There have been suggestions that the sunspot cycles having a period of about 11 years can also contribute to these variations. This has, however, been disputed by many researchers.

We proceed to look at the available data, more closely.

It has become possible to use powerful methods of mass-spectroscopy to infer the average temperature of the Earth and the extent of carbon dioxide in its atmosphere over the last 800,000 years.

The European Project for Ice Coring in Antarctica drilled ice from Dome C in Antarctica reaching up to a depth of 3270.2 meters, just 5 meters above the bedrock. The carbon dioxide content is easily measured, as air bubbles are trapped in the ice.

The temperature is inferred by measuring the deuterium and oxygen-18 content of the ice sample and the fossil marine shells, respectively.

Hydrogen has two stable isotopes, H (hydrogen, having one proton and no neutrons) and D (deuterium, having one proton and one neutron). Oxygen has three stable isotopes ^{16}O (having eight protons and eight neutrons), ^{17}O (having eight protons and nine neutrons), and ^{18}O (having eight protons and ten neutrons). Thus, water molecule can have compositions: H_2O, HDO, and D_2O, where oxygen could be either ^{16}O, or ^{17}O, or ^{18}O. The chemical properties of these are identical. However, their physical properties differ as deuterium is heavier than hydrogen. Similarly, ^{18}O is heavier than ^{17}O, which is heavier than ^{16}O. The natural abundance of deuterium is 0.0115%, while those of ^{16}O, ^{17}O and ^{18}O are 96.76%, 0.04%, and 0.20% respectively.

Because of the increasing mass, H_2O will evaporate more easily than HDO, which will evaporate more easily than D_2O. Similarly, $H_2{}^{16}O$ will evaporate more easily than $H_2{}^{18}O$, *etc.* These rates are quite sensitive to temperature.

When water from the oceans evaporates, the vapor has a smaller abundance of deuterium than the water in the ocean. It will form a cloud and come back to the poles as snow which will become ice with passage of time. More snow accumulates above it, registering the temperature of the atmosphere for posterity. The D/H helps us to calculate the temperature with a great accuracy.

Similarly, oxygen (contained in water) can get incorporated in calcium carbonate to form the shells of marine animals. If the ocean temperature is lower, the abundance of ^{18}O would be higher than normal, as more of ^{16}O leaves the ocean by evaporation. Thus, a measurement of ^{18}O in fossil marine shells gives an independent measurement of the temperature. The two measurements are found to agree, leading to a great deal of confidence in the procedure. Similar results have been obtained by measurements of ice core from Greenland and at a different point in Antarctica.

The results of the measurement of the temperature, provide two important results. First, they reveal a cyclic pattern of brief warm spells (see the peaks in the graphs) followed by long spells of cold, called interglacial. The rise to higher temperatures is rapid whereas the drop to cold periods is slow. Secondly and more interestingly, the pattern repeats with a period of approximately 100,000 years.

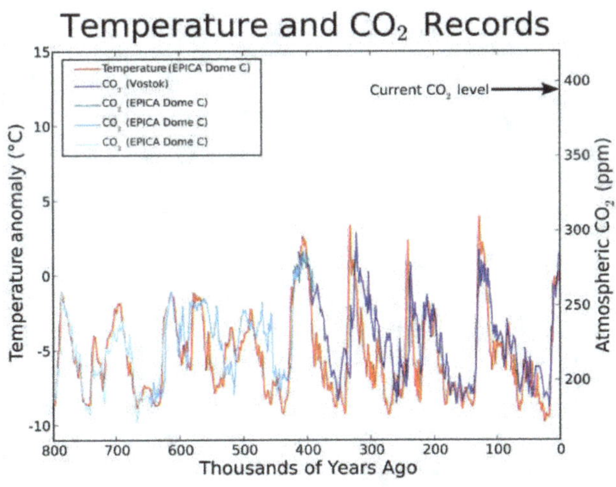

Fig. 6.4: Climate of The Planet Earth for the Past 800,000 Years [68]

A comparison of the temperature profile with the variation in the carbon dioxide results for the carbon dioxide concentration strongly suggests that an increased concentration of carbon dioxide in the atmosphere leads to higher temperatures. A very important observation is that in the entire period of 800,000 years before 1800 AD — the beginning of industrialization, the level of carbon dioxide has never been more than 300 ppm. We shall have a closer look at this, a little later and first look at the periodicity of the rise and fall of the temperature in the past.

It is of interest to recall that as early as in 1896, Svante August Arrhenius had suggested that the increase in carbon dioxide in the atmosphere could lead to increase in average temperature of the Earth, due to the greenhouse effect.

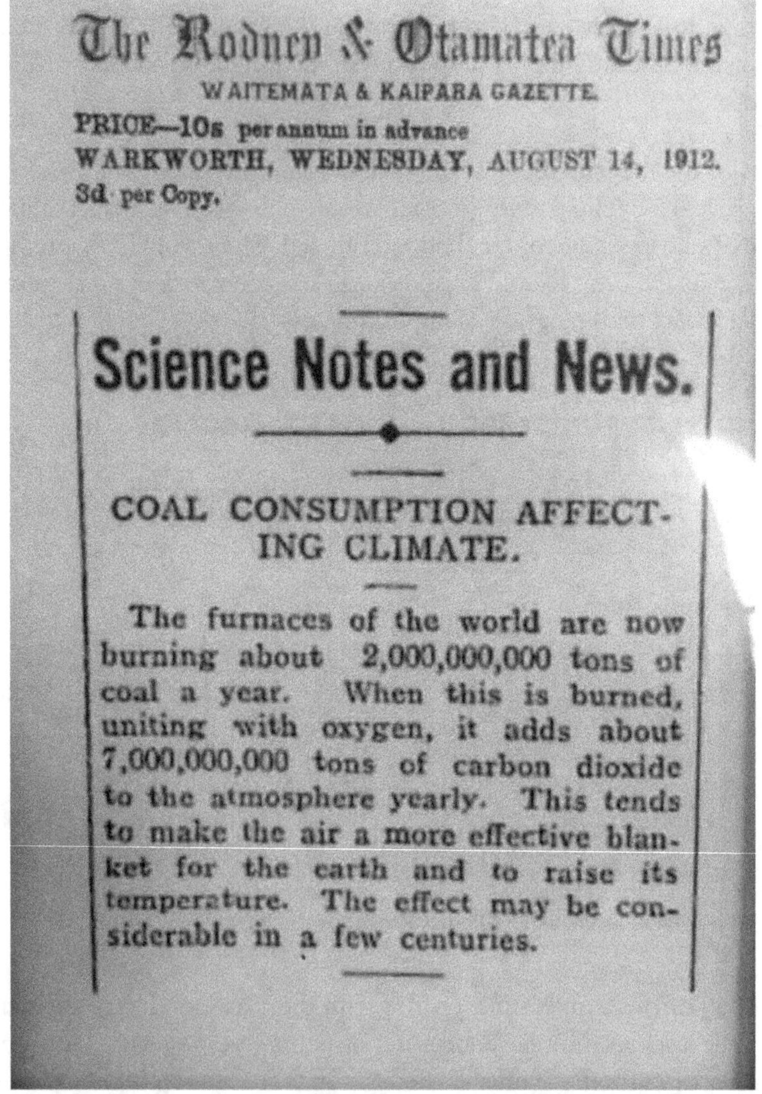

The Rodney & Otamatea Times

WAITEMATA & KAIPARA GAZETTE.

PRICE—10s per annum in advance

WARKWORTH, WEDNESDAY, AUGUST 14, 1912.

3d per Copy.

Science Notes and News.

COAL CONSUMPTION AFFECTING CLIMATE.

The furnaces of the world are now burning about 2,000,000,000 tons of coal a year. When this is burned, uniting with oxygen, it adds about 7,000,000,000 tons of carbon dioxide to the atmosphere yearly. This tends to make the air a more effective blanket for the earth and to raise its temperature. The effect may be considerable in a few centuries.

Fig. 6.5: A News Item on Warming of Atmosphere Published in 1912 [69]

However, as Europe was going through a spell of cold at the time of his predictions, it was even considered beneficial! It is also interesting to recall a news item from 1912, which raised this issue.

Most of the discussion of greenhouse gases centres around carbon dioxide in the atmosphere. Even though, a large part of the carbon dioxide produced by us is absorbed by the oceans (where it leads to other complications discussed earlier), the remaining part can stay in the atmosphere for hundreds or even thousands of years. We shall not discuss water vapour, as it has a lifetime of only a few hours or days in the atmosphere, as it is removed by rains or snowfall.

Now let us look at the temperature and carbon dioxide concentration in recent past.

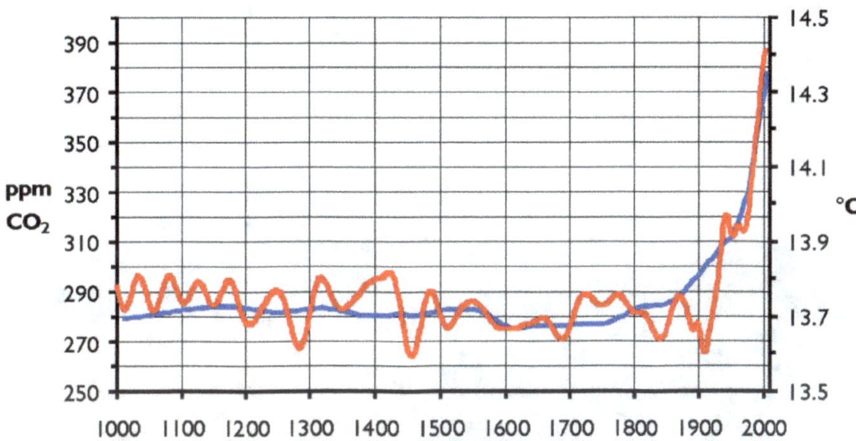

Fig. 6.6: Temperature (in red, scale on right) of The Earth and Concentration of Carbon Dioxide (in blue, scale on left) [70]

A very clear correlation between the rising global temperatures and the concentration of the carbon dioxide is seen. As mentioned earlier, till 1800 or so, the concentration of carbon dioxide in the atmosphere never exceeded 300 ppm. It has risen monotonically to about 415 ppm in 2019 from 280 ppm in 1800.

The concentration of other major greenhouse gases due to human activities is also rising. To have a quantitative discussion on the relative importance of the greenhouse gases, the concept of Global Warming Potential (GWP) has been developed. GWP is a measure of the heat trapped by a greenhouse in the atmosphere up to a specific time horizon, relative to carbon dioxide of the same amount. It depends on the absorption of infrared radiation by the given species, the spectral location of its absorbing wavelengths, and its atmospheric lifetime.

Table 6.1: Global Warming Potential of some Greenhouse Gases [71]

Green House Gas	Global Warming Potential for 100 Years
CO_2	1
CH_4	23
N2O	296
HFC-23	12,000
HFC-134a	1,300
SF_6	22,000

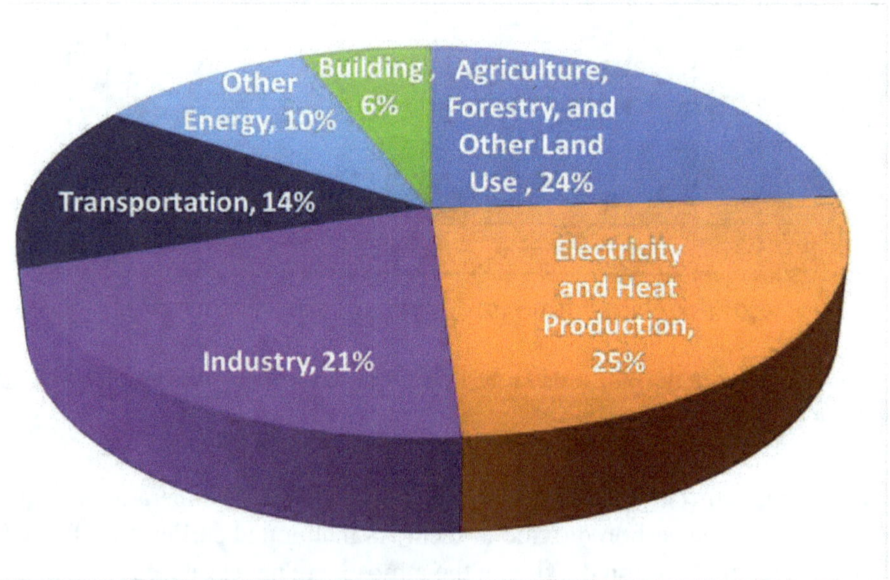

Fig. 6.7: Global Greenhouse Gas Emissions by Economic Sector for the Year 2010 [72]

These numbers must be used along with the concentration of these gases in the atmosphere.

We give the contribution of individual industries (Fig. 6.7).

Typical values of the relative contributions of the gases can be seen in the pie chart given below, for a typical year of 2010 across the World, based on the report of the Intergovernmental Panel on Climate Change 2014.

Here we shall discuss only two of the other greenhouse gases, methane and nitrous oxide.

Methane

We have seen that the next major source of greenhouse gases is methane. Methane is a potent greenhouse gas which has a lifespan of about 12 years in the atmosphere, as it is removed by chemical reactions. Kilogramme for kilogramme, it is more than 20 times as effective as carbon dioxide in trapping heat and is responsible for 16% of the global warming, while carbon dioxide is responsible for 76% of it.

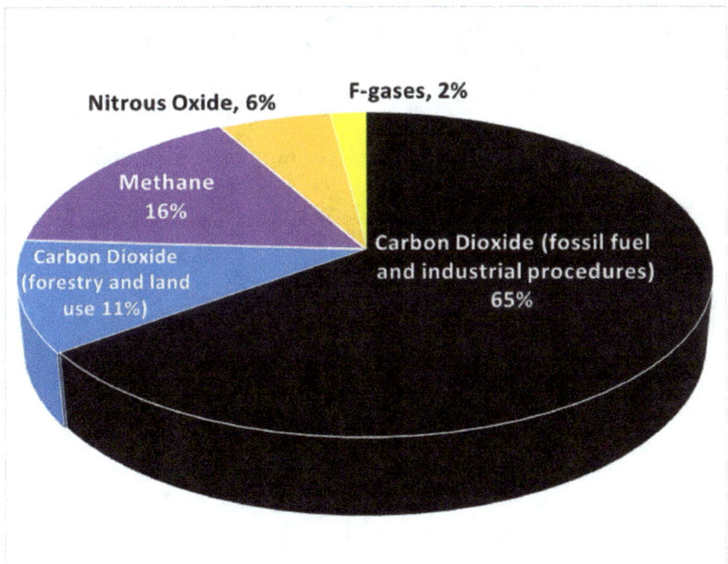

Fig. 6.8: Global Greenhouse Gas Emission by Gas for the Year 2010 [73]

Methane is produced by mining of fossil fuels, rotting vegetation, landfills, city wastes, paddy fields and digestive system of animals. A potentially enormous source of methane is permafrost and methane hydrate (found under sediments on the ocean floor of the Earth), which may be released by global warming.

The concentration of methane in the atmosphere and its close correspondence with the temperature over more than 400,000 years were obtained from the measurement of ice core drilled at Vostok, Antarctica.

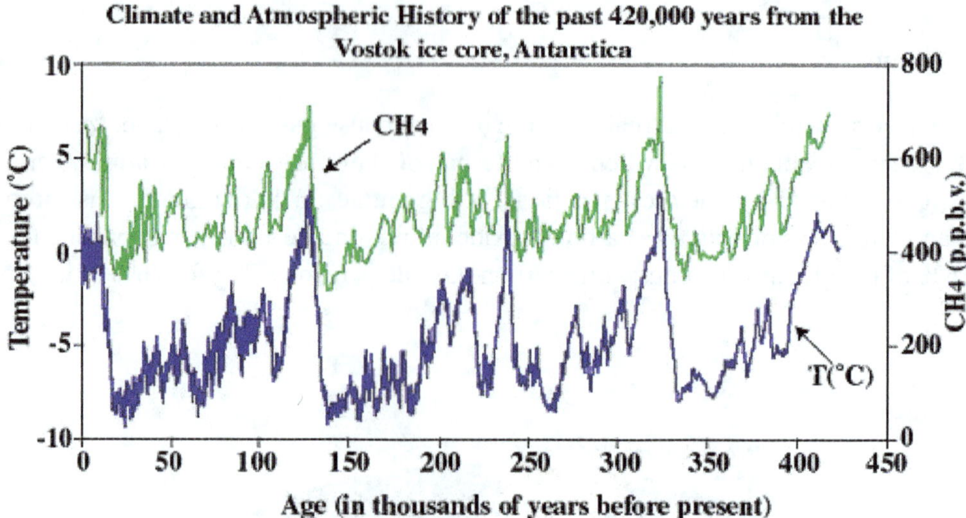

Fig. 6.9: Concentration of Methane and Global Temperature [74]

A close correspondence in rising temperatures and concentration of methane is seen. The rise of methane in recent times is alarming, and in fact, it has more than doubled in recent times compared to its average value in the past.

The remaining 6% of the global warming is attributed to nitrous oxide. Nitrous oxide is not only a greenhouse gas, but also responsible for potentially hazardous depletion of ozone at ground level as well as in the upper atmosphere, in addition to having harmful effects on humans.

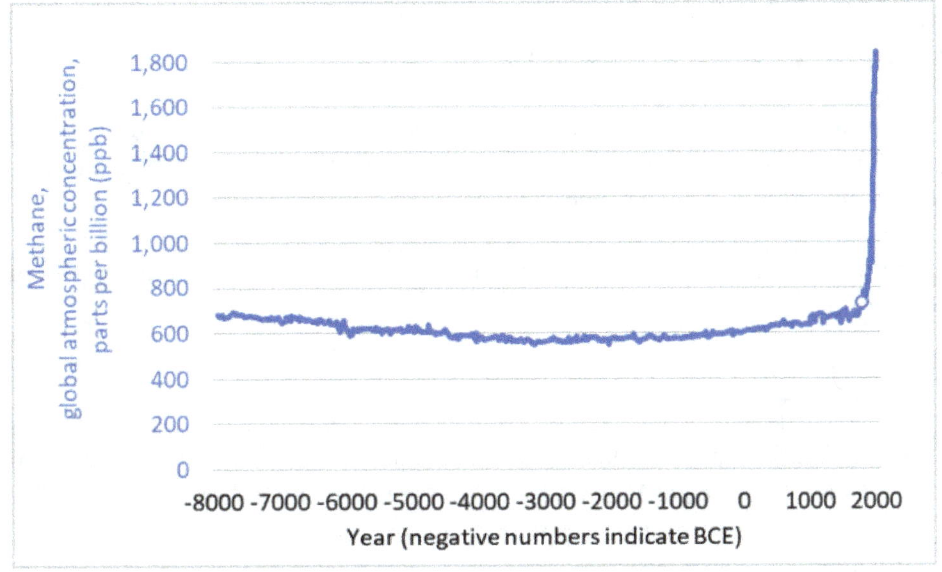

Fig. 6.10: Concentration of Methane [75]

Nitrous Oxide

Although N_2O makes up a relatively small amount of the atmosphere (about 330 parts per billion parts of atmosphere), it has a GWP of about 300 and atmospheric lifetime of about 120 years.

Nitrous oxide is produced by both natural and human sources. The major natural sources include soils under natural vegetation, tundra, and the oceans. Important human sources come from agriculture (nitrogen-based fertilizers, soil cultivation), livestock manure, biomass or fossil fuels combustion, and industrial processes. In total, about 40% of the emissions are estimated to be due to human actions. The nitrous oxide levels are now higher than at any other time during the last 800,000 years. Since the industrial revolution, the atmospheric concentrations of nitrous oxide have increased by about 20%.

Alarmed by these results, the Intergovernmental Panel on Climate Change (IPCC) has been publishing reports on the climate change data and recommendations, which were issued in 1990, 1995, 2001, 2007, and 2014. The next report is expected in 2022.

Those reports provide a vast body of detailed and careful analysis of global warming, its signs, and its consequences.

These establish beyond any doubt that "*Climate change is occurring, is caused largely by human activities, and poses significant risks for — and in many cases is already affecting — a broad range of human and natural systems*".

It is important that we look at the summary of the reports of the three working groups of the 2014 report:

Working Group I

- "Warming of the climate system is unequivocal, and since the 1950s, many of the observed changes are unprecedented over decades to millennia".
- "Atmospheric concentrations of carbon dioxide, methane, and nitrous oxide have increased to levels unprecedented in at least the last 800,000 years".
- "Human influence on the climate system is clear. It is extremely likely (95–100% probability) that human influence was the dominant cause of global warming between 1951–2010".

Working Group II

- "Increasing magnitudes of [global] warming increase the likelihood of severe, pervasive, and irreversible impacts".
- "A first step towards adaptation to future climate change is reducing vulnerability and exposure to present climate variability".
- "The overall risks of climate change impacts can be reduced by limiting the rate and magnitude of climate change".

Working Group III

- "Without new policies to mitigate climate change, projections suggest an increase in global mean temperature in 2100 of 3.7 to 4.8 degree Celsius, relative to pre-industrial levels (median values; the range is 2.5 to 7.8 degree Celsius including climate uncertainty)".

- "The current trajectory of global greenhouse gas emissions is not consistent with limiting global warming to below 1.5 or 2 degree Celsius, relative to pre-industrial levels. Pledges made as part of the Cancún Agreements are broadly consistent with cost-effective scenarios that give a "likely" chance (66–100% probability) of limiting global warming (in 2100) to below 3 degree Celsius, relative to pre-industrial levels".

The most important recommendation that IPCC makes is to make every possible effort to try to limit the increase in global temperature to less than 1.5 or at most 2 degrees Celsius of those of pre-industrial limits by every possible means.

This requires that the net carbon dioxide emissions decline by about 45% from the 2010 levels by 2030 and reach net zero around 2050.

Chapter 7

Energy and Human Development

Development is the natural aspiration of every human being — every individual would like to have a 'better' tomorrow than today, every individual will like his children to have 'better' tomorrow than they themselves had. However, the word 'better' is relatively undefined with different meanings to different people. Ever since man discovered the use of fire around a million years ago, his development has been rapid. Cooking made more foods edible year-round, released more nutrients and calories from both vegetables and meat, and killed harmful bacteria, viruses, and parasites. This led to increase in the brain size of early humans and gave him extra time to develop problem solving skills. Since then, man has not looked back. It is often said that man's development was driven by fire and his brain.

The most important desire, enshrined in Sanskrit literature suggests:

सर्वे भवन्तु सुखिनः, सर्वे सन्तु निरामयाः ।
सर्वे भद्राणि पश्यन्तु, मा कश्चिद्दुःखभाग्भवेत् ॥
शान्तिः शान्तिः शान्तिः ॥

("*May all be happy. May all be free from illness. May all see what is auspicious. May no one suffer. May this bring Peace to individuals, Peace to the Nation, and Peace to the World*").

This desire is obviously too holistic to be true with the ground reality being very different. The strong–weak divide, the rich–poor divide, *etc.*, have always been a part of human civilization. The technology rich–technology poor divide is the new paradigm of the twenty first century. How do we ensure reasonable level of development for every human being?

A quantitative measure of human development is however required for a deeper understanding of issues of human development across the world. A statistic composite, Human Development Index (HDI), has been suggested by Mahbub ul Haq and further refined by a score of authors. It has become a very useful measure of the development of a country. The following definition of HDI has been adopted by United Nations since 2010.

- A long and healthy life, measured as the Life Expectancy Index at birth

  Life Expectancy Index (LEI) is defined as [(LE-20)/(85-20)]. Thus, LEI is 1 if the life expectancy at birth is 85 years and zero if the life expectancy at birth is 20 years. One may recall that the life expectancy in India in 1901 was close to 24 years and in 1947 when India became independent it was still quite low, just about 37 years. In 2019 it is close to 69 years, which is still lower than 85 years or more, in developed nations.

- Mean years of schooling and expected years of schooling, measured in terms of Education Index

  Educational Index (EI) is defined as the arithmetic mean of Mean Years of Schooling Index (MYSI) and Expected Years of Schooling Index (EYSI). MYSI is defined as MYS/15, implying an average 15 years of schooling for our youth. EYSI is defined as EYS/18, as in most of the countries in the world, a master's degree requires 18 years of education.

- A decent standard of living measured in terms of Gross National Income converted to international dollars using purchasing power parity.

  The Income Index (II) is defined as [{ln(GNIPC)-ln(100)}/{ln(75000)-ln(100)}]. Thus, it is one if the Gross National Income Per Capita is $75,000 and is zero if it is $100.

Finally, the Human Development Index is obtained by taking a geometric mean of the three indices defined above:

$$\mathrm{HDI} = (\mathrm{LEI} \times \mathrm{EI} \times \mathrm{II})^{1/3}$$

HDI values lie between zero and one; one would imply a perfectly happy and zero would imply a completely miserable country.

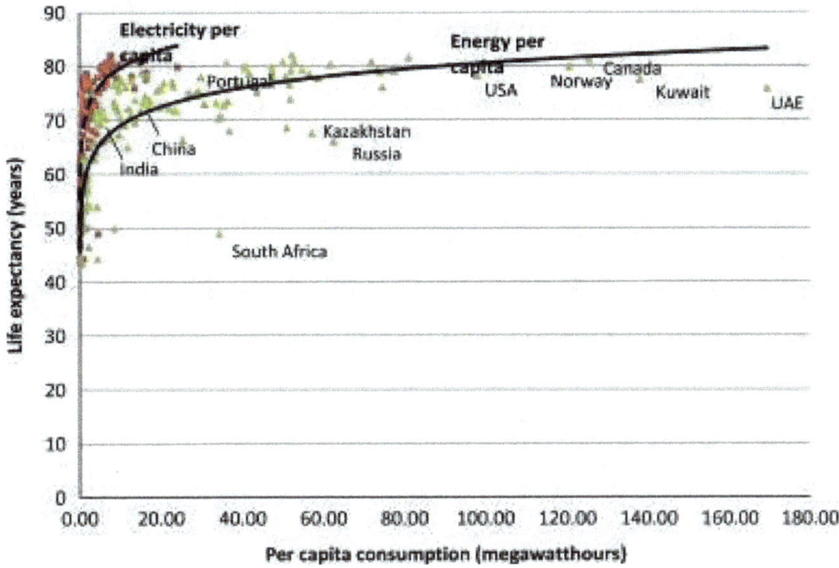

Fig. 7.1: Life Expectancy Rises with Availability of Power/Electricity [76]

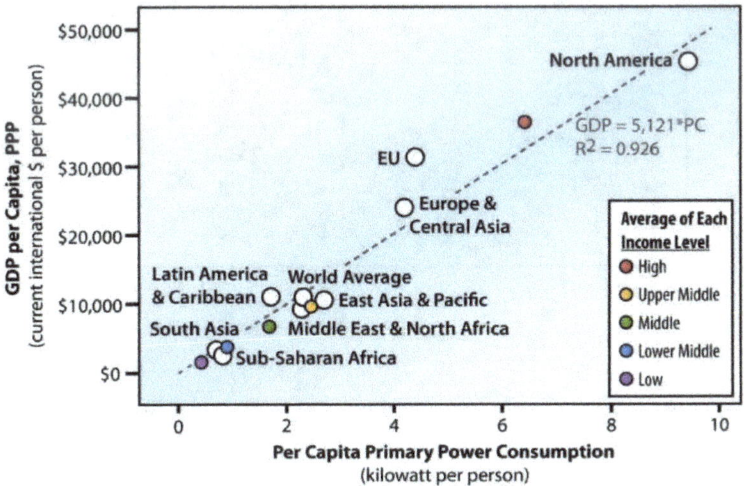

Fig. 7.2: The Gross Domestic Product Rises with Increased Availability of Power [77]

It is known that life expectancy increases with the utilization of more energy or electricity across the nations of the world. Similarly, Per Capita Gross National Income or Gross Domestic Product increases with Per capita Primary Power consumption.

The world presently has values of HDI varying from about 0.354 for Niger to 0.953 for Norway. India has HDI of 0.64 and China 0.752 in 2019. Almost eighty percent of the world population (including about 800 million people in China) has an HDI less than 0.8, while the industrialized world has HDI of mostly more than 0.8. This study reveals that a power consumption of the order of 3000–4000 kWh/person/year is enough to have HDI of the order of 0.8, which is considered a good life. It is also seen that the developed nations use as much as 10000 kWh/person/year or even more. India's per capita per year electricity consumption is about 1181 kWh.

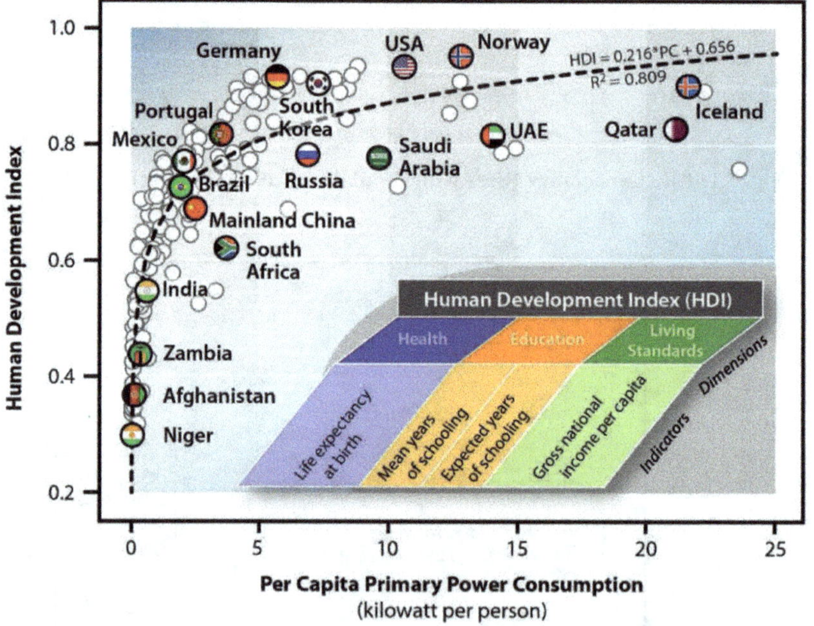

Fig. 7.3: Human Development Index and Per Capita Power Consumption [77, 78]

There should be no doubt that human well-being is decided by the rate at which power is consumed, or the rate at which work is done. Even, though now put on a firm quantitative footing, this relationship should not surprise us.

Let us discuss this in a little more detail. Over an 8-hour work shift, an average, healthy, well fed, and motivated manual labourer may sustain an output of around 75 watts of power or 0.6 kWh of energy in a day. He can enhance this a few times using simple tools like levers and many more times by using more complex tools or animals like oxen or horses. Till the onset of the Industrial Revolution, a power usage of the order of 3000 kWh per year was confined to only some members of the aristocracy or extremely rich merchants who had access to many men and beasts under their control. For the first time in history, the living standards of ordinary people underwent an unprecedented and sustained growth, during the Industrial Revolution. This has been possible as the world has used fossil fuels to provide energy for these developments. It is generally believed that if the world does not maintain this high rate of consumption, the results discussed above suggest that we would get poor, sick, and less educated.

However, countries outside the developed part of the world like India and African nations are yet to reach *even* this minimum rate of energy consumption and an acceptable Human Development Index. With about 85% of the global energy consumption coming from fossil fuels like coal and oil, we already have a disaster in the making. Not only the world resources of fossil fuels are unlikely to last more than 100 years even at the present rate of consumption, but it will also cause irreparable and irreversible damage to the environment and to the wellbeing of the world population.

We have no alternative but to adopt a rational mix of energy resources, fossil fuels and renewable energy resources, in order to provide enough energy to all people of the world to raise them to a level where their Human Development Index is close to 0.8 without damaging the environment.

The amount of energy needed for this to happen is about 30 trillion kWh per year, double the present amount. This would require that we need to consider all the alternatives, including the sustainable and green sources of energy for our survival immediately.

Chapter 8

How Much Energy Do We Need?

Before we embark on a description of various sources of energy and possible options available to us for our sustainable future, let us get an idea about the enormity of this task by estimating the energy requirement of the world in coming years based on estimates of population and their aspirations.

We estimate the above, based on estimates of population, increase in GDP, and projected improvement in technology. We give these projections for energy and its contribution to carbon dioxide in the table to follow.

The world energy consumption rate, (DE/dt), is obtained by multiplying the world population (N), per capita GDP of the world, and energy consumption rate per unit of GDP. This way of writing brings out most clearly the increases and decreases in the rate of energy consumption as its contributing factors increase or decrease.

The world population was about 6.1 billion in 2001 and is projected to increase to 9.4 billion by 2050 assuming a reasonable rate of increase of 0.9%/year. Per capita GDP of the world was about $7,500 in 2001, and may increase to about $15,000 by 2050, assuming a rate of increase of 1.4%/year.

World energy intensity was approximately 0.29 Watts/$/year in 2001 and it can be projected to decrease by 0.8%/year to approximately 0.20 Watts/$/year by 2050.

These projections in the fundamental factors provide that the world energy consumption rate is expected to more than double from 13.5 Terra Watt in 2001 to 27.6 Terra Watt in 2050 and then to more than triple to 43 Terra Watt by 2100.

As a sobering thought, the total emission of atmospheric carbon dioxide is given in the last column. To do this, we first estimate carbon emission by multiplying the rate of consumption of world energy (doe/dt) with the amount of emission of carbon per unit energy (C/E, or carbon intensity). The emission of carbon dioxide can then be estimated easily.

We add that the carbon intensity for fossil fuels are 0.78 kick/(Watt.Year) for coal, 0.60 kegs/(Watt.Year) for oil, and 0.47 kgs./(Watt.Year) for natural gas. The projections in the table are much lower than those of any of the fossil fuels. The world carbon intensity has been assumed as 0.41 kgC/(Watt.Year). This assumes that a significant fraction of the world's energy would be supplied either by non-fossil fuels, or by fossil fuels with carbon sequestration. If this is not so, the numbers above will have to be correspondingly revised. We realise that if the carbon intensity decreases to 0.24 kgC/(Watt.Year) by 2050 and to 0.14 kgC/(Watt.Year) by 2100, the increased use of energy will not increase the emissions of carbon (and carbon dioxide) beyond the level of 2001. This brings out an urgent need to find clean sources of energy while meeting the increasing demand of energy.

Table 8.1: Energy Demand Based on World Energy Statistics and Projections [79]

Quantity	**Definition**	**Units**	**2001**	**2050**	**2100**
N	**Population**	**Billion**	6.145	9.4	10.4
GDP	Gross Domestic Product	Tera $/year	46	140	284
GDP/N	Per Capita GDP	$/(person.year)	7,470	14,850	27,320
[dE/dt]/GDP	Energy Intensity	Watt/($/Year)	0.294	0.20	0.15
dE/dt	Energy Consumption Rate	Terra Watt	13.5	27.6	43.0
C/E	Carbon Intensity	Kilogramme (kg) C/(Watt.Year)	0.49	0.41	0.30
dC/dt	Carbon Emission Rate	Giga ton (Gt) C/Year	6.57	11.0	13.3
d(CO_2)/dt	Equivalent CO_2 Emission Rate	Giga-ton (Gt) CO_2/Year	24.07	40.03	48.8

However, it should be emphasized that the carbon emission rate of 6.6 GtC/Year in the year 2001 was already too high.

Atmospheric models which consider the absorption of the emitted carbon dioxide by oceans have revealed that the carbon emission rate of 1–2 GtC/Year would have stabilized the atmospheric carbon dioxide to 550 ppm raising it from the pre-industrial level of 280 ppm. If we retain the level of carbon emission to 6.6 GtC/Year, the carbon dioxide level would rise to 1470 ppm.

The emission of carbon dioxide in 2019 was about 37 Gt, which implies a carbon emission of about 10 Gt.

With this sobering realization of the enormity of the task of providing enough energy for our masses to pull them out of poverty without smothering our planet with global warming, we proceed to discuss various sources of energy.

Chapter 9

Energy Resources

We have already pointed out that the only source of energy that the early humans had was the Sun and the muscle power of himself and of animals. Over a period, he learnt to tame the power of fire, blowing wind, and running water. The discoveries of coal and oil were major milestones in human history which not only triggered the Industrial Revolution of the eighteenth century but also promised to power the industries for a long time to come. But the growth of the energy usage over the recent decades tell a different story. While the global energy usage has grown by leaps and bounds and continues to grow with hydrocarbon fuels — coal, oil, and natural gas, forming the backbone of our energy resources, it has also been realized that the present and the projected rates of consumption of these resources cannot be maintained for ever. Some estimates indicate that the world will run out of these resources in less than a century. It was also mentioned earlier that continued burning of hydrocarbons have started loading the Earth's atmosphere with more and more carbon dioxide, resulting in an appreciable increase in global temperature with catastrophic consequences. It is not surprising that the world has started relooking at some of the traditional sources of energies like wind and solar power that are not only renewable but are also environment friendly. More renewable energy resources like tidal and geothermal energies have also been identified in the recent years. Recent developments in semiconductor technology have opened the possibility of drawing electrical energy directly from the sunlight using photovoltaic panels. Twentieth century also saw the emergence of nuclear energy as an option to satisfy partly, the growing energy needs of the world.

In this and the following chapters, we give a brief description of the various sources of energy along with their strengths and weaknesses. For obvious reasons, renewable energy resources like solar, wind, hydro, tidal and geothermal are also referred to as "Inexhaustible Resources". We shall discuss these one by one, except for Uranium and Thorium, which will be discussed in separate chapters on Nuclear Power to follow. Nuclear power based on uranium and thorium are generally listed under non-renewable sources of energy. However, their energy density is so high that the known sources may last us for thousands of years.

9.1 Coal, Oil, and Natural Gas

Coal is the backbone of the world's electricity supply today providing more than 40% of our electricity needs through thermal power stations. It is also the earliest and the most widely available fossil fuel resource. Additionally, coal is a key component in the production of steel and concrete, the two basic building materials. It has been projected that coal will continue to provide a major fraction of the 'on-grid' electricity for quite some time.

As we have mentioned earlier, coal was made when giant plants, perhaps predating even the dinosaurs, died in swamps, got buried under water and dirt for millions of years, while pressure and heat converted them into coal.

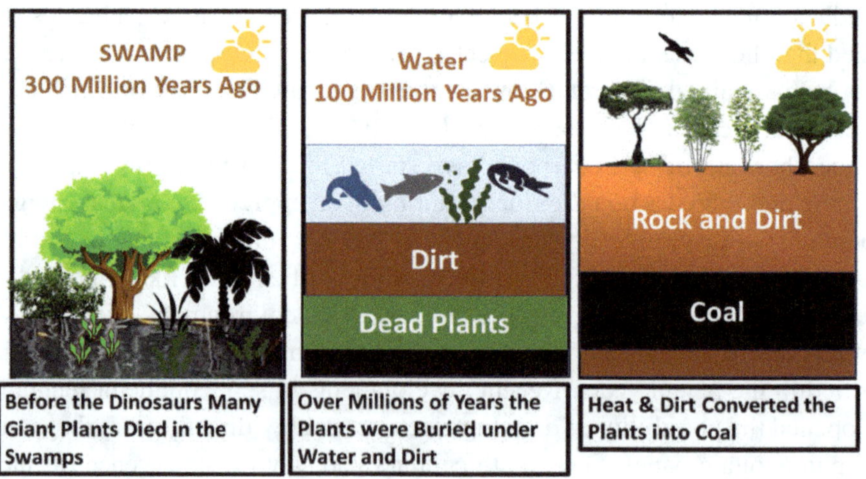

Fig. 9.1: Formation of Coal [80]

Coal mining provides more than several million jobs worldwide. At the same time, it has also been recognized as a very hazardous activity, killing more than 15,000 people every year, with many more involved in serious mining accidents. One of the well-known hazards faced by the miners is carbon monoxide build up in the mines. Till the advent of robust electronic sensors of carbon monoxide (an odourless and colourless poisonous gas) in the eighties, the miners used to descend into the mines accompanied by a canary. Canaries are very sensitive to carbon monoxide and other toxic gases and were used as sentinels to warn the minors about the impending danger. "Canary in a coal mine" has since become a phrase for all such sentinels. Till the introduction of Humphrey Davy's lamp in 1815, they were also victims of eruption of flames due to release of inflammable gases like methane.

Fig. 9.2: A Coal Miner with a Canary [81]

The miners and people living around mines are exposed to a variety of poisonings, leading to pre-mature deaths, and the growth of children is severely affected. There are several coal mines across the world, where coal seam fires have been razing for decades, releasing enormous amount of poisonous gases and particulate matter in the atmosphere. One such mine at Jharia, in Jharkhand has a fire razing since 1916. There are reports that the particulate matter released by such coal mines and thermal power plants, when deposited on ears of grain or fruits, while growing, destroys them. A classic example is the development of holes in cheekus (sapota) from Dahanu in Maharashtra, due to deposition of fly ash on the fruits in their nascent stages.

Fig. 9.3: Coal Miners

The coal also needs to be carried from pitheads or ports to thermal power stations, over long distances. These coal trains are known to pollute the entire route with coal dust. The deposit of coal dust severely impacts the fertility of agricultural land.

Fig. 9.4: Coal Wagon Trains

Fig. 9.5: A Coal Mine in Dhanbad [82]

It is known that coal mining severely alters the landscape, which destroys the natural environment of the surrounding land. Coal mining eliminates existing vegetation, destroys the genetic soil profile, displaces, or destroys wildlife and habitat, alters current land uses, and permanently changes the general topography of the area mined. The dust released in the process of removal and stacking of topsoil severally affects the health of the people living near mines. Mine collapses are frequent and endanger the life and property of people living nearby. World Health Organisation estimates suggest that up to eight million premature deaths every year can be directly attributed to ill consequences of using fossil fuels and burning of biomass. In fact, coal mines are often called as "cancer of the soil".

Oil and Natural Gas

Next only to coal, oil has become the world's most important source of energy in the modern world. It is the fuel of choice for all transport vehicles, by road, by air, or by sea. Oil is also the backbone of industries manufacturing fertilisers, plastics, detergents, and even medicines.

Oil and natural gas were formed when tiny sea plants and animals died and were buried on the ocean floor.

As time passed these were covered by multiple layers of silt and sand. These remains got buried deeper and deeper and the enormous heat and pressure turned them into oil and gas. We must drill down layers of sand, silt, and rock formations to get to oil and natural gas.

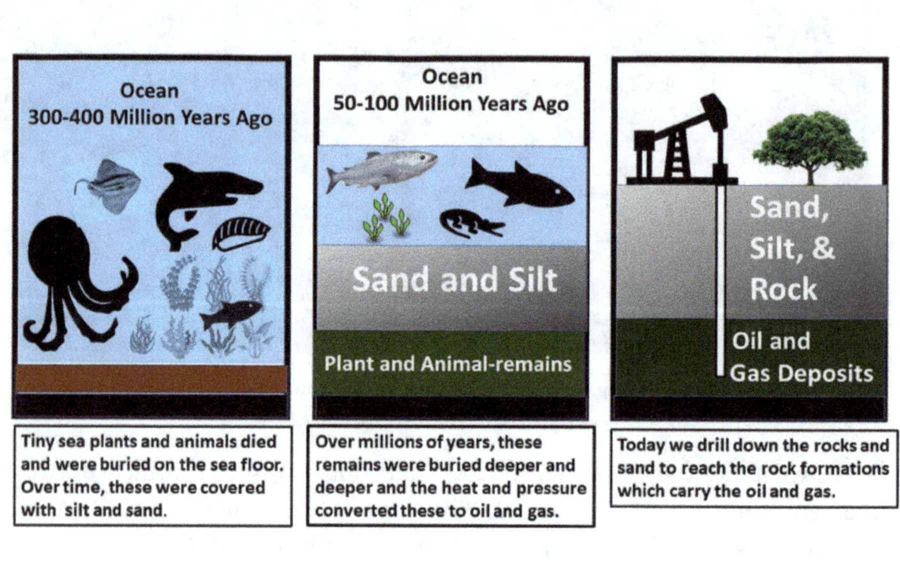

Fig. 9.6: Formation of Oil and Natural Gas [83]

One of the serious hazards of oil drilling is oil spills. Oil spills were known to have taken place in various countries of the Mediterranean, Asia Minor, and the Far East since the ancient times. Herodotus, Plutarch, Pliny the Elder, and Marcus Vitruvius described the occurrences of oil-spills in various parts of India, Persia, Syria, Mediterranean islands, Malaysia, and China, *etc.* Herodotus also wrote that bitumen from a small tributary of the River Euphrates (in modern-day Iraq), was used for the construction of the wall of Babylon, the most famous city of ancient Mesopotamia. Sumerians used bitumen to make roads and for waterproofing their boats. Indus Valley people used it to waterproof the Great Bath and their baskets for storing grain. There is evidence of the use of bitumen in Ancient Egypt. Chinese are known to have transported natural gas to their homes using bamboo pipes for heating and to cook meals, centuries ago.

Alexander the Great, while campaigning in the fourth century BC on the southern shores of the Caspian Sea and the River Ochus (today's Azerbaijan) was intrigued by the bright yellow lights being lit using "a yellow liquid that flowed through a tube". The natives called this light "Chirak" and the fluid was known as "Naphtha", which means "the liquid leaking from the ground".

Now, of course, this oil is being extensively utilized, since the first oil well was drilled by Edwin L. Drake in Pennsylvania in 1859. Two severe environmental effects of oil and natural gas extraction should be mentioned here. Even though, transporting crude oil is easier and cheaper than transporting coal, disastrous oil tanker spills are quite common.

Oil spills coat everything they touch and severely affect the ecosystem which they enter. When an oil slick from a large spill reaches a beach, the oil coats and clings to every rock and grain of sand. When the oil washes into coastal marshes, mangrove forests, or other wetlands, fibrous plants and grasses absorb oil, plants are damaged, and the area becomes unsuitable for habitation of wildlife.

Fig. 9.7: Environmental Effects of Oil Spill [84]

When oil eventually stops floating on the water's surface and begins to sink into the marine environment, it has similar damaging effects on fragile underwater ecosystems, killing or contaminating fish and smaller organisms that are essential links in the global food chain.

Almost all natural-gas wells burn a part of the gas. Additionally, there could be accidental fires at these wells.

This burning natural gas releases gases like sulphur dioxide and carbon dioxide, which lead to acid rains.

Fig. 9.8: Burning Gas Well

Fig. 9.9: A Forest After an Acid Rain

Acid rains destroy trees and severally damage monuments made of marble and sandstone.

Fig. 9.10: Acid Rain Breaks Down Calcium Carbonate in Sandstone and Marble [85]

Additional oil is often extracted from these wells by forcing carbon dioxide or nitrogen gases into the well to push oil out.

Natural Gas

Like coal and oil, natural gas is yet another fossil fuel formed deep below the surface of the Earth millions of years ago when the remains of plants and animals broke down under pressure and heat. The decaying matter formed gas and got trapped in porous rocks in areas that were later covered by harder rocks. Natural gas is primarily made of methane, the lightest hydrocarbon. It burns more cleanly than other fossil fuels emitting lower levels of harmful emission such as carbon monoxide, carbon dioxide and nitrous oxides and produces less greenhouse gases than coal and oil. It does not produce ash or particulates that cause health problems. It is not as clean as renewable energy sources such as solar and wind but could function as a stop gap measure as society makes the transition to clean and renewable energy sources.

To feed the world's demand for natural gas, it needs to be transported from where it is produced to where it is going to be consumed. Transporting gas is not easy. LNG is created by cooling natural gas to −160 degrees C creating a clear colorless non-toxic liquid 600 times denser than natural gas.

Shale Oil and Shale Gas

Shale oil is an unconventional oil extracted from oil shale rock fragments, organic-rich fine-grained sedimentary rock containing kerogen. Kerogen is a solid mixture of organic chemical compounds. Shale oil is often used as a substitute for conventional crude oil. Extracting shale oil is costly, both financially and in terms of its environmental impact. These include issues such as large-scale land use, waste management, and air and water pollution. In addition, the combustion and thermal processing of the shale oil generate hazardous waste material and harmful atmospheric emissions, including carbon dioxide. As shale is mostly mined from the surface, it causes large environmental impact associated with open-pit mining and converts vast tracks of land to waste, in addition to releasing harmful rock dust into the atmosphere. It also releases harmful gases including carbon dioxide. More than 1,000 people have already been killed in USA alone during extraction of shale oil and gas.

Fracking, which involves forcing water or some other liquid under high pressure to break-open the rocks holding the natural gas or oil is also used to extract shale oil or gas. As an aside, it might interest the readers to note that this liquid is often made by adding water to gum of guvar beans (string beans), which used to be supplied by India. Now most of it is supplied by Brazil. Fracking has serious environmental consequences as it may contaminate the water table. It is also known to cause land subsidence and repeated tremors.

Fig. 9.11: Semi-coke Heap, Spent Shale, and Open Pit Mines left after Shale Oil/Gas Extraction [86]

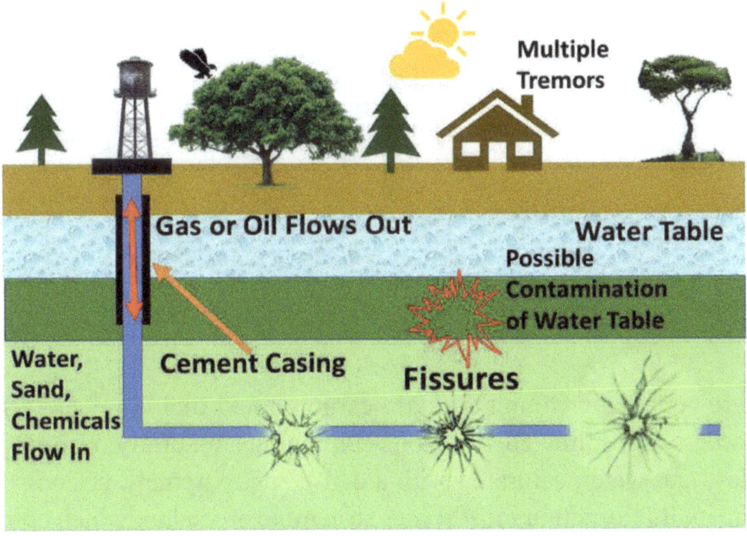

Fig. 9.12: Fracking for Extraction of Oil or Gas [87]

It is estimated that India may have 3 trillion cubic meters of Shale Gas. This could be extracted if we can solve the technical difficulties, and the environmental aspects are addressed. Some studies, however, suggest that the shale in some of the likely rich areas in India has a lot more clay, from where it may be difficult to extract oil and gas as compared to rocks in USA and Canada.

Fig. 9.13: Shale Gas Deposits in India [88]

Oil Sand

Oil sands are yet another source of hydrocarbon liquid fuel. They are either loose sands or partially consolidated sandstone containing a naturally occurring mixture of sand, clay, and water, saturated with a dense and extremely viscous petroleum called bitumen. Its extraction mostly uses strip mine procedure which converts vast stretches of forests to waste. These are also close to major rivers and severely affect the water bodies.

Some countries, especially Canada, Kazakhstan, Russia, and Venezuela have vast deposits of oil sands. As with the extraction and use of any fossil fuel, harmful and negative effects arise because of the extraction, upgrading and processing of bitumen from the oil sands. These mainly include:

- Impact of Tailing Ponds: Tailing ponds are settling ponds that contain the waste by-product of oil sands extraction and upgrading. They are a mix of water, sand, silt, clay, unrecovered hydrocarbons, and other contaminants. The large concentration of these chemicals makes reclamation of the affected area very difficult.

- Impacts on the Climate: The greenhouse gas emissions from oil sand extraction and processing are significantly larger than that for conventional crude oil. These emissions contribute to global warming and the enhanced greenhouse effect.

- Impact on Water: The extraction of bitumen from oil sands requires a large amount of water, and thus water use is a concern when looking at oil sands extraction. In principle, water used in the oil sands can be recycled, but only small amounts of this water are returned to the natural cycle.

- Impacts on Quality of Air: Along with greenhouse gases, other pollutants including oxides of nitrogen and sulphur are released in the atmosphere.

Coal, oil, and natural gas are considered non-renewable because they do not get replenished in a short period of time.

Electricity from Fossil Fuels

We have already mentioned that while fossil fuels are abundant sources of energy, the most common form of energy from the point of view of transport and use is electricity. Electricity generation using fossil fuels is straight forward — burn the fuel and generate heat, use the heat to generate steam, and use the steam to operate a turbine and generate electricity. Electricity can be transported easily over long distances. A schematic diagram of a typical thermal power station is shown in the figure below.

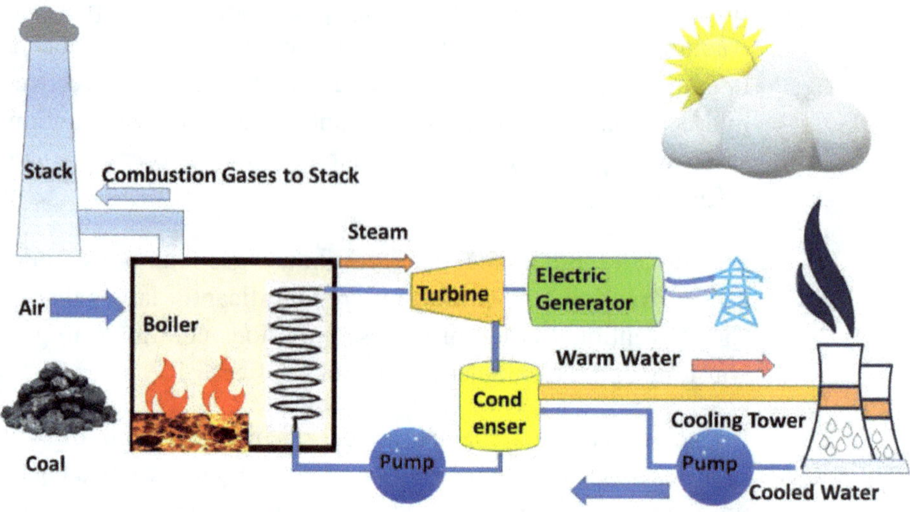

Fig. 9.14: Schematic Diagram of a Typical Coal Fired Thermal Power Plant [89]

We first note that thermal power plants using natural gas, which are less polluting than coal-fired plants are also more efficient. We also note that the efficiency of gas power plants can be further improved by using combined cycles where the power plant uses both a gas and a steam turbine together to produce up to 50

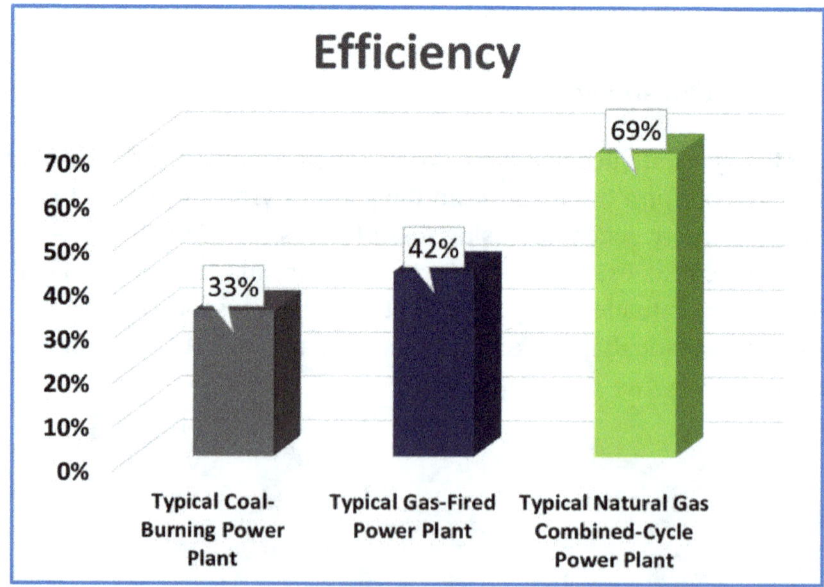

Fig. 9.15: Efficiency of use of Coal, Oil and Natural Gas [90]

percent more electricity from the same fuel than a traditional simple-cycle plant. The waste heat from the gas turbine is routed to the nearby steam turbine, which generates extra power. The efficiencies of power plants using biomass are generally lower, with efficiencies as low as 22%.

Before closing this discussion of fossil fuels as energy resources, let us also have a brief look at the various steps in their use along with the efficiency of their use, which is quite instructive as it points to steps where more efficient procedures need to be adopted.

We see that if we use an incandescent lamp to get light from coal, losses at successive steps — in generating electricity, in transmitting, and then conversion to light leads to an overall efficiency of just 1.2%.

Use of compact fluorescent lamps or light emitting diode lamps can improve this efficiency by a factor of four to five, by producing more light from the same amount of electricity and therefore of coal. We do need to intensify our research efforts into developing more efficient technologies for more efficient utilization of the fuel. This discussion is included here only to illustrate the point that it is possible to make improvements which can reduce our energy requirement and therefore the fuel requirement.

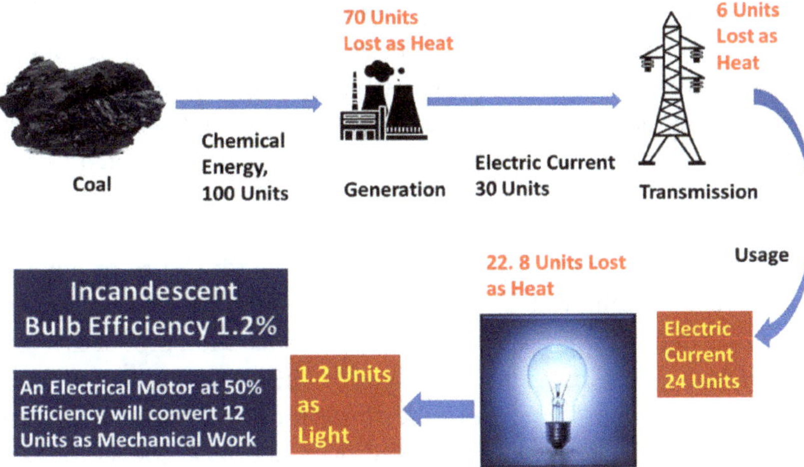

Fig. 9.16: Losing Energy at Every Step! [91]

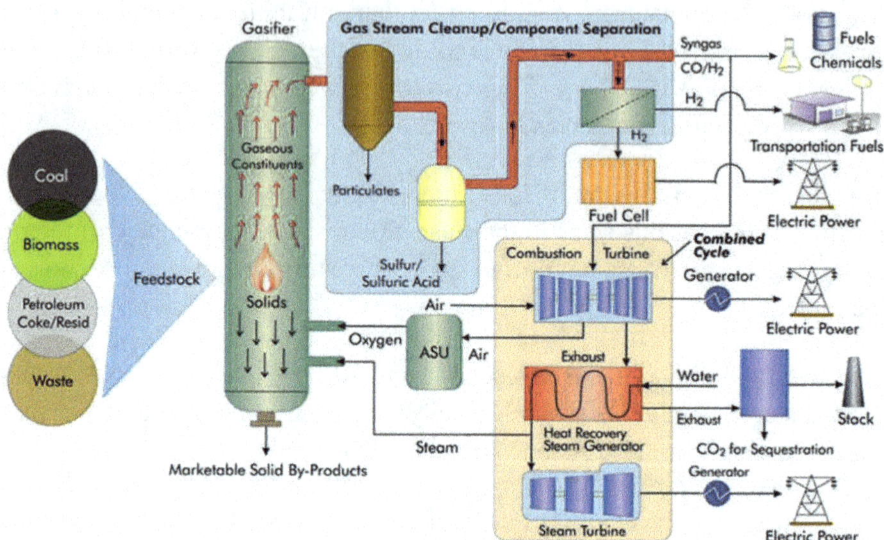

Fig. 9.17: Scheme for Coal Gasification [92]

Before closing this discussion on Coal and Natural Gas, we add that coal gasification, which was earlier used to produce the so-called "town" gas to provide street lighting has reached a high level of sophistication. The large availability of coal (and other carbon-based materials like biomass and city waste) across the world makes it a very attractive proposition, *provided* the capture of the resulting carbon dioxide for further use and production of hydrogen for use in fertilizer and chemical industry as well as fuel are fully implemented. The other pollutants like sulphur dioxide and ash, *etc.*, can also be easily separated, in this process. Carbon capture, however, is not yet deployed extensively and a lot more research and development are necessary for its wide application. It is also expected to be quite expensive.

The process uses coal to produce synthesis gas (syngas for short), or town gas as mentioned above. It is achieved by reaction of coal, air (oxygen) and steam in a high temperature, high pressure vessel. The chemical reaction produces a mixture of carbon monoxide and hydrogen. The syngas can be converted to hydrogen and carbon dioxide using what is known as the shift process. It involves its reaction with steam in the presence of a catalyst. Hydrogen, as we know, is the cleanest source of energy known to mankind. It is possible in principle to capture carbon dioxide from syngas, which prevents its emission as a greenhouse gas. It can be used for enhanced oil extraction or it can be kept in safe storage.

The coal gasification scheme can be adopted to convert coal into syngas from deep mines. China is planning to implement this scheme on a large scale. However, utilization of this scheme in deep mines may lead to land subsistence and cause serious accidents.

In absence of a complete and permanent capture of carbon dioxide, the process would be a massive source of carbon dioxide.

We close this discussion by recalling that coal was used for decades to power steam engines and engines used in railways. In early days, even trams of many cities used coal for its engines. We have seen that the invention of internal combustion engines of varying powers revolutionised transport, and now ships, railways, airplanes, cars, trucks, and other automobiles run on oil. In fact, about 15% of the greenhouse gases in the atmosphere are contributed by the fossil fuels used for transportation.

9.2 Biomass and Biofuel

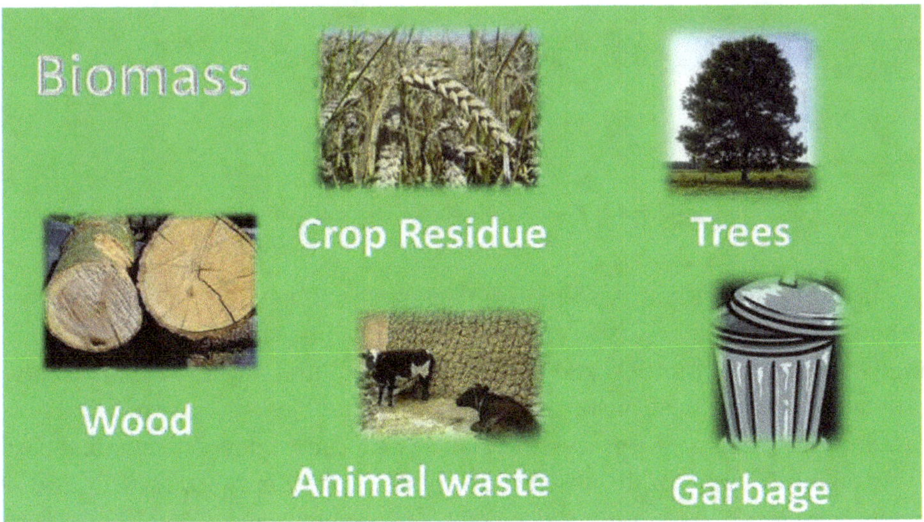

Fig. 9.18: Biomass [93]

Plants and animals have always been contributing to usable energy resources to humanity since time immemorial. For example, wood, agricultural waste, forest produce, fruits, nuts as well as human and animal wastes, all contribute to the available energy resource pool to mankind.

One should also remember that the energy contained in biomass originally came from the Sun through photosynthesis, where carbon dioxide in the air is converted to sugars, starches, and cellulose. Biomass is therefore a renewable energy resource.

$$6\ CO_2 + 6\ H_2O + \text{Energy} \rightarrow C_6H_{12}O_6 + 6O_2$$

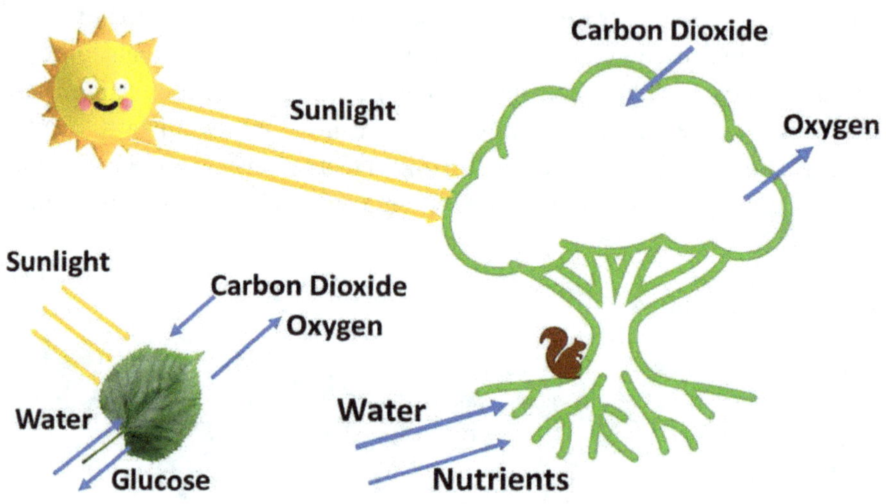

Fig. 9.19: Photosynthesis for Bioenergy [94]

The chemical energy that is stored in plants and animals (animals eat plants or other animals) or in their waste is also referred to as bioenergy. Crop residues form an important part of biofuels and include all agricultural wastes such as bagasse, straw, stem, stalk, leaves, husk, shell, peel, pulp, stubble, *etc.* Rice produces both straw and rice husks at the processing plant which can be conveniently and easily converted into energy. In the absence of suitable opportunity for their utilization, farmers have resorted to burning rice stubbles — which causes untold miseries to their neighbourhood. It is rather ironic that rice plants which use up to 2000 litres of water for a kilogramme of rice, along with fertilizers, and pesticides end up in such a way, producing particulate matter too in the process, which takes days or weeks to clear. There have been indications of black snow at places as far as Darjeeling due to burning of these stubbles in Punjab, Haryana, and Western Uttar Pradesh.

Fig. 9.20: Burning of Rice Stubbles

Significant quantities of biomass remain in the fields in the form of cob when maize is harvested which can be converted into energy. Sugar cane harvesting leads to harvest residues in the fields, while fibrous bagasse is produced when sugar cane is crushed to extract juice. Vast quantities of coconut husk are produced from harvesting of coconuts.

Animals and poultry produce a great deal of manure. Fruit processing industries produce a great deal of peelings and pulp. This along with waste from food processing industry, municipal wastes, sewage, and waste from timber industry provide a large amount of biomass, which can all be burned to produce energy, in principle.

It is estimated that processing 1,000 kg of wood in a sawmill produces 520 kg of waste. Further, processing of 1,000 kg of wood in the furniture industries leads to a waste generation of 450 kg.

There is one problem though; and it is not a trivial problem. Agricultural wastes are seasonal.

Fig. 9.21: Bagasse from Sugar Cane Industry

In principle, biomass is just a product of solar energy. The plants capture carbon dioxide from the atmosphere and convert it to biomass. Burning the biomass releases the carbon dioxide back into the atmosphere. Thus, utilization of biomass does not lead to any net increase of carbon dioxide in the atmosphere.

We should remember though that a large tree may have absorbed the carbon dioxide for over hundred years. A rice plant may have absorbed it over several months. When we burn these, we release the resulting carbon dioxide in a few minutes. Thus, there would be a spike in the concentrations of carbon dioxide (and particulate matter — if not trapped), which however will ultimately average out over time. This is in addition to the importance of the trees for providing us oxygen and as controller of moisture.

One other problem with biomass, especially agricultural and wood waste is that their density is very low, and these are spread over large areas. Thus, their collection and transport to plants using them is labour intensive and expensive. On-site compactification of the waste can be quite useful, though expensive.

Fig. 9.22: Coconut Husk

The biomass, especially the animal waste can be used as a fertilizer, putting the organic carbon and nitrogen back into soil. This age-old practice has come under strain as now the animals and poultry are often given special feeds, which are fortified with growth hormones and antibiotics, etc., and their usage as fertilizer has been found to introduce these into our food chain. It has also been reported that several plants using biomass tend to add chippings of used tyres to the biomass to attain higher temperatures during ignition. This can lead to emission of toxic gases.

An important indicator of the availability of biomass is forest-cover of the world.

It is ironic that with passage of time, rich nations have increased their forest cover, while poorer nations have seen a drastic decrease in their forest cover. It is seen that high-income countries keep pushing the deforestation problem to the poorer parts of the world. Heavily indebted poor countries are especially affected.

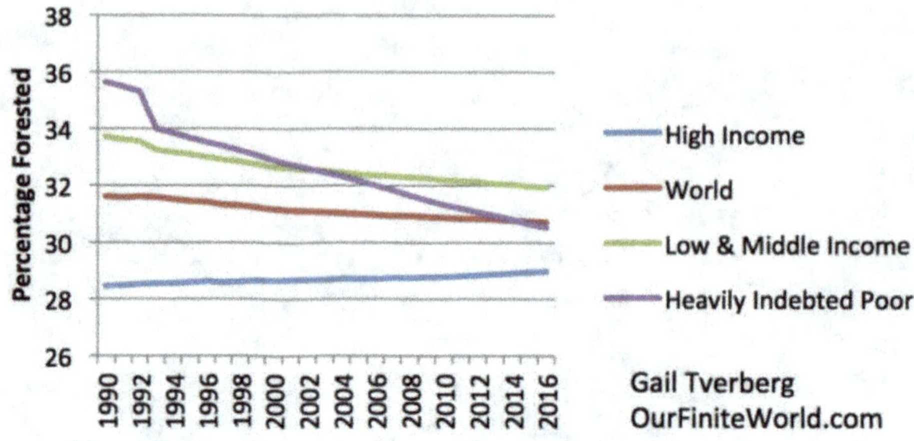

Fig. 9.23: Forestation for the Rich and Poor Nations [95]

Biofuels

Biofuels, e.g., ethanol and biodiesel are being used by several countries, especially USA and Brazil, as a supplement to petrol or diesel or as a replacement for these. These were promoted to reduce dependence on import of petroleum.

Cellulosic materials from plants, e.g., wood, stems, and leaves can be converted into sugars by physical and chemical processes and using enzymes. The sugar can then be converted to ethanol or other liquid biofuels by fermentation or chemical processes.

While it is generally advisable to leave agricultural residues in the fields to replenish the organic carbon, nitrogen, phosphorus, and other trace elements, plants like poplar and switch grass, *etc.*, can be grown on land not suitable for agriculture.

Fig. 9.24: Biofuel [96]

Of late, however, several countries have started using corn, sugar cane, sugar beet, sorghum, barley, potatoes, sweet potatoes, cassava, sunflower, fruits, and wheat, *etc.*, for this, which seriously affects the availability of food and prices for food. USA, for example, diverts more than 40% of its corn production for production of ethanol. Several of these crops usually require a great deal of water for irrigation.

Fig. 9.25: Food vs. Fuel [97]

It has also been argued that North America and Europe have a large surplus of food production, which brings down the prices of food grains, leading to loss for their farmers. This prompts them to divert these resources for production of biofuels. The resulting rise in food prices will seriously affect the poorer nations.

To produce palm oil, several countries like Indonesia and Malaysia have been clearing large tracks of their rain forests with their rich biodiversity to cultivate just oil palms. It is estimated that 1,000 to 5,000 orangutans are killed every year as their habitat is lost. The damage to ecosystem due to the clearing of rain forests is irreparable.

Biodiesel

Biodiesel is generally produced from rapeseed, soybean, mustard, sunflower, flax, palm, and coconut oil, *etc.* It is also produced from waste vegetable oil and animal fats. Again, there have been discussions about this route diverting food supplies for oil production.

The major sources of feedstock (raw material) for making biodiesel in the United States and their shares of total biodiesel feedstocks in 2018 were

- soybean oil 54%
- corn oil 15%
- recycled feedstocks (such as used cooking oils and yellow grease) 13%
- canola oil 9%
- animal fats 9%

This led to production of 6.3 billion Kg of biodiesel.

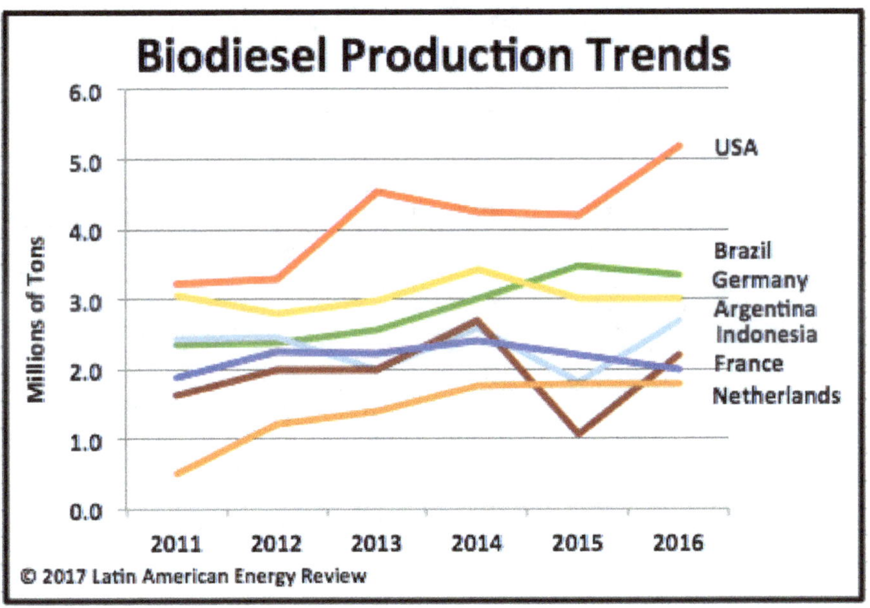

Fig. 9.26: Biodiesel Production [98]

Rapeseed oil, sunflower oil, and palm oil are major feedstocks for biodiesel produced in other countries.

Biodiesel is normally blended with petroleum diesel in ratios of 2% (referred to as B2), 5% (B5), or 20% (B20). Pure biodiesel (B100) can also be used in many applications. Diesel engines can use biodiesel fuels without changes to the engines. Biodiesel blends are also used as heating oil.

However, it should be mentioned that production of biofuels is also a very energy- and chemical-intensive process. It also remains a very small component of the energy used by the world.

The CO_2 Impact of Coal, Oil, Natural Gas and Biofuels

We have already pointed out the various environmental effects of usage of coal, oil, natural gas, and biofuels for energy. It is worthwhile having a quantitative idea about their contribution to atmospheric carbon dioxide load.

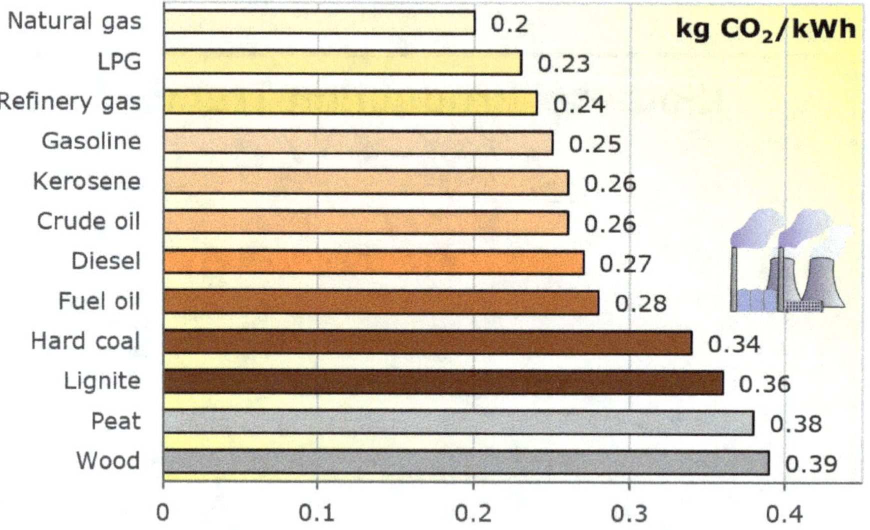

Fig. 9.27: Carbon dioxide Emission per kWh of Power Produced [99]

The carbon dioxide emission of any fuel is decided by the carbon content of its molecules. Thus, coal has a larger carbon content than the oil and natural gas.

Specific carbon-dioxide emission of various fuels in terms of kilogramme of carbon dioxide emitted per unit kWh is shown in the figure above. We see that use of natural gases and oil in thermal power plants does reduce the emission of carbon dioxide considerably.

The enormity of the problems can be gauged by looking at the following table which gives various pollutants per year by a 1300 MW thermal power plant, using coal. It is expected to use 4.6 million tons of coal per year.

Table 9.1: Pollutants Emitted by a 1300 MW Thermal Power Plant using Coal

Pollutant	Quantity
Carbon dioxide	14.5 Million Tons
Ash	1.3 Million Tons
Gypsum	650,000 Tons
Sludge	27,000 Tons
Nitrous oxide	38,000 Tons
Sulphur dioxide	23,000 Tons
Particulate matter	1,300 Tons
Uranium	6.5 Tons

Use of washed coal, which removes ash, soil and rocks from the coal can reduce the production of fly ash from thermal power plants.

Electrostatic precipitators and scrubbers can, in principle, successfully remove up to 90 percent or more of several of the above pollutants, especially sulphur dioxide. However, they are very costly to install and operate and increase the cost of electricity by 25% or much more. Retrofitting of these facilities, which require space and addition to the exhaust chimneys in existing plants is not always easy.

Carbon capture and storage, where waste carbon dioxide is captured from a large point source, e.g., a cement factory or a thermal power plant and transported to a storage site and deposited in an underground geological formation is also possible but quite expensive. It is also not yet known whether the trapped carbon dioxide will not leak back from such deposits. A lot more research and technological development is necessary before it is widely and economically deployed.

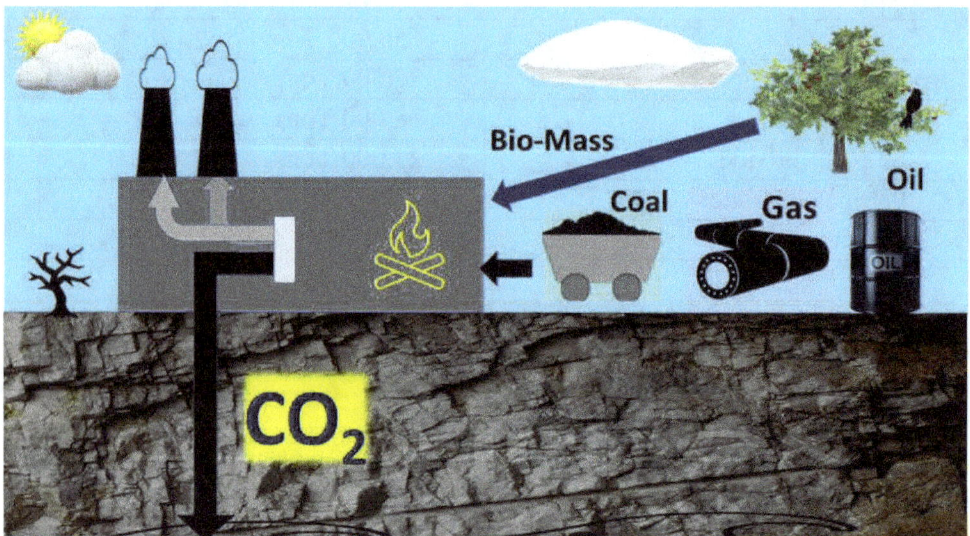

Fig. 9.28: Principle of Carbon Capture and Storage

9.3 Wind Energy

We have already mentioned that wind and flowing water have always been valuable sources of energy to human beings. We have also seen that sail boats have been in vogue for thousands of years for intercontinental migration and trade. Similarly, windmills have been in vogue for a variety of jobs like grinding corn or lifting water for centuries.

Fig. 9.29: A Windmill in Kinderdijk, The Netherlands (Built in 1740) [100]

Uneven heating of the atmosphere by the Sun, rotation of the Earth, variations in the Earth's topography such as the presence of mountains, bodies of water and vegetation in its path, all contribute to strong winds that can move things.

It has been estimated that wind energy can provide a power of up to 250 Watt/m^2 of the area swapped by its rotors.

In the recent decades, an increasing number of windmills are being built to generate electricity and use it for stand-alone applications or feed into the grid. However, it should be remembered that the energy generated depends on several parameters specific to the location. A small change in the velocity of the wind leads to a large change in the energy production by the windmill. The produced energy is also proportional to density of the air which depends sensitively on the site elevation and the temperature.

Fig. 9.30: Windmills with Horizontal and Vertical Axes [101]

Considerable progress has also been made in the design and development of windmills in the recent times. Detailed maps of wind energy potential have also been mapped across the potential sites.

By now, both horizontal and vertical axis windmills having rotors with diameters of a few hundred metres and generating several MW of power on a windy day are commercially available. Sea shore as well as offshore and even floating installations have been perfected.

Even though they need a reasonably large initial investment, windmills and windmill farms have several advantages.

Wind power is one of the cleanest sources of energy with no greenhouse gas emission, except at the time of their manufacture and installation. We have already noted that it is a renewable source of energy. It does not need any fuel and only occasional maintenance. These essentially do not occupy any space, as the rotors are at a height of 50 metres or more.

On the other hand, these can be a source of continuous noise pollution and there are reports of birds, especially bats, being killed due to their rotating blades. A surprising outcome of this is increase in lizard population at several windmill stations.

The greatest problem with windmills is the excessive variability of the wind and hence fluctuations in the power generated. If the wind stops, the production drops to zero.

Fig. 9.31: Paddy Fields and Wind Turbines, India [102]

The requirement of steady power for industry and homes would then necessitate using windmills along with thermal or nuclear power stations to cover the base load and which will have to be operated at varying capacities, often much below their full capacity, which puts a financial strain on them. Even though thermal power plants can be ramped up or ramped down at the rate of about 1% per one to five minutes, this adds to the cost of electricity production. Some studies have even suggested that running thermal power stations with fluctuating production leads to more greenhouse gases. It has been suggested that some form of energy storage like large capacity batteries or production and storage of hydrogen can take care of this supply demand mismatch. Urgent work is required to address this issue.

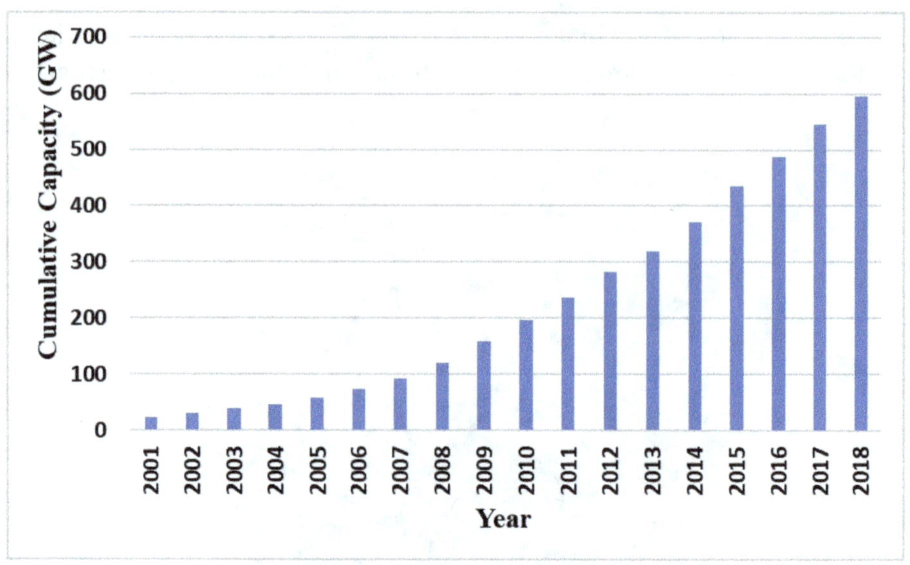

Fig. 9.32: Evolution of Global Wind Energy Installed Capacity in GW [103]

Considering all of the above, the installed capacity of wind power across the world is continuously on the rise. It is expected that wind power could reach 2,000 GW by 2030 and supply up to 17–19% of global electricity. This should create more than 2 million new jobs and reduce CO_2 emissions by more than 3 billion tons per year by 2050. In several countries, the excess energy produced by windmills is being used to produce hydrogen — the greenest source of energy, by electrolysis of water.

9.4 Geothermal Energy

We have seen that the inner and outer cores as well as the mantel of Earth have large dimensions and are very hot. The heat content of Earth is estimated to be 2.8×10^{15} TWh. It flows to the surface of the Earth at the rate of 44.2 TW. Radioactive decay of Potassium-40, Thorium-232, Uranium-235, and Uranium-238 replenishes this to the extent of 30 TW. This energy is more than twice the energy requirement of the world today. However, it is too diffuse and on the surface of the Earth, it amounts to about 0.1 Watt/m^2. Recall that the solar power at the surface of the Earth can be as high as 1000 Watt/m^2. There is a temperature gradient of the order of 20–30 degree Celsius per kilometre as we go deeper below the Earth's surface. Local variations can bring these hot regions closer to the surface.

Fig. 9.33: Snow Monkeys of Japan Taking Advantage of Hot Springs to Stay Warm [104]

Fig. 9.34: Bison Keeping Warm During Winter Near a Steaming Hot Spring [105]

Water passing through these hot regions will be heated and may even be converted to steam. It can come to the surface as geysers or hot springs, known from prehistorical times. Not only humans, even animals, e.g., snow monkeys of Japan, are known to take advantage of such hot springs.

The bison in the Yellowstone region of USA are known to keep warm by staying close to hot springs during winter — the water is too hot for a dip though.

Hot springs usually exist in volcanic regions or in areas where there are extensive faults. The water circulates through the fault zones (basically damaged zones with high permeability) bringing heat from great depths. However, these faults do not have to be active. Hot springs may continuously produce hot water at moderate to high temperatures.

Geysers are hot springs that intermittently spout a column of hot water and steam into the air. This is caused by the water in deep conduits beneath a geyser which reaches the boiling point.

Utilisation of geothermal energy is relatively straight forward. Of course, if we are close to a hot spring with voluminous flow one simply drills a hole to the hot aquifer. Else, one can create a hot aquifer by drilling a hole to push cold water down and another one to get hot water (if only heating is needed) or steam out, which can then be used for heating or generation of electricity. The aquifer is created by breaking the rocks between the two holes by fracking.

Obviously, deeper holes/wells will reach hotter zones, unless of course a hot thermal reservoir is closer to surface. While one or two-kilometre-deep holes are common, wells as deep as 12 kilometres have become possible with technological advances.

The ease with which the geothermal energy can be tapped has prompted many countries to use it for heating their homes. Iceland uses geothermal energy on a very large scale.

Geothermal power plants also need only about 400 square metres of land for one GWh compared to more than 3600 square metres for coal power plants and 1300 square metres for windmill and solar farms. They use just about 20 litres of fresh water per MWh compared to 1000 litres for nuclear, coal, and gas plants.

Fig. 9.35: A Geyser in Haukadalur, Iceland

Geothermal energy has some emissions of sulphur dioxide, carbon dioxide, hydrogen sulphide, and ammonia gases and arsenic, mercury, boron, and antimony. Thus, care must be taken that these emissions can be filtered, and this water does not contaminate the surface water. It has been seen that existing thermal power plants emit up to 122 kilogrammes of carbon dioxide per MWh of electricity, which is a minor fraction of what plants using fossil fuels produce, though.

If the geothermal energy is used only for heating houses etc., we would still need to run the pumps using normal electricity, whose contribution to greenhouse gases should not be ignored. Often, especially if the source of the heat is closer to the surface of the Earth — it gets exhausted after some years necessitating a move to some other location. There have been instances of land subsidence when geothermal energy was tapped.

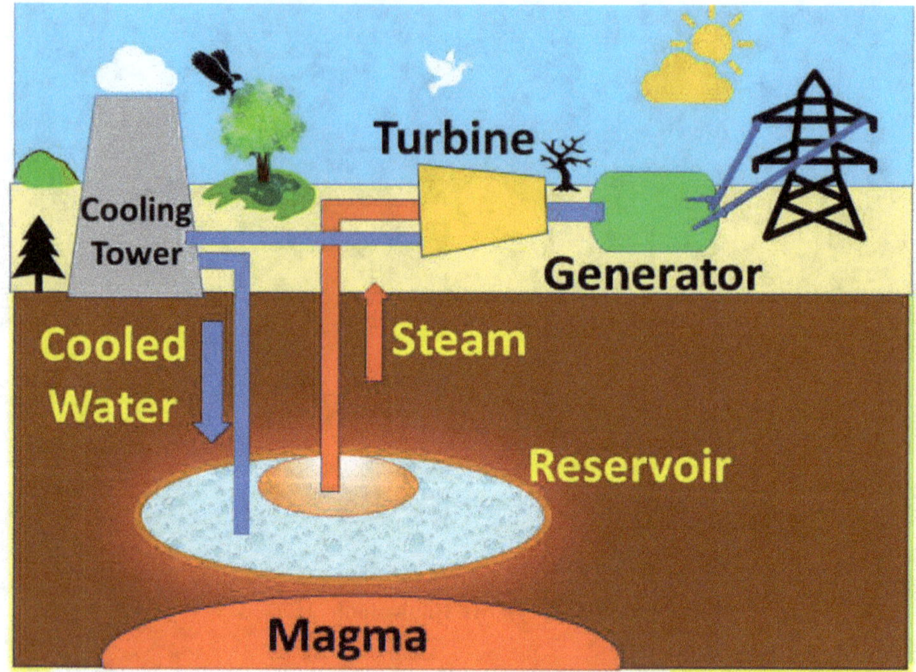

Fig. 9.36: The Principle of using Geothermal Energy [106]

Most of the commercial-grade production of geothermal energy is located along "geothermal systems", where the heat flow is closer to the surface of the Earth and hot water or steam can rise either to the surface, or lies at a depth that we can easily reach by drilling. Many of these regions occur within the so-called "ring of fire", a ring of geothermal sites.

By early twenty-first century, geothermal energy was being used to produce electricity in more than twenty countries. The main users have been United States, Philippines, Indonesia, Mexico, New Zealand, Italy, and Iceland. The installed capacity for geothermal energy across the world is now close to 15,000 MW.

The important geothermal areas in India include the Himalayas, Sohana, West Coast, Cambay, Son-Narmada-Tapti (SONATA), Godavari, and Mahanadi.

The geothermal potential of India is estimated to be about 10,000 MW.

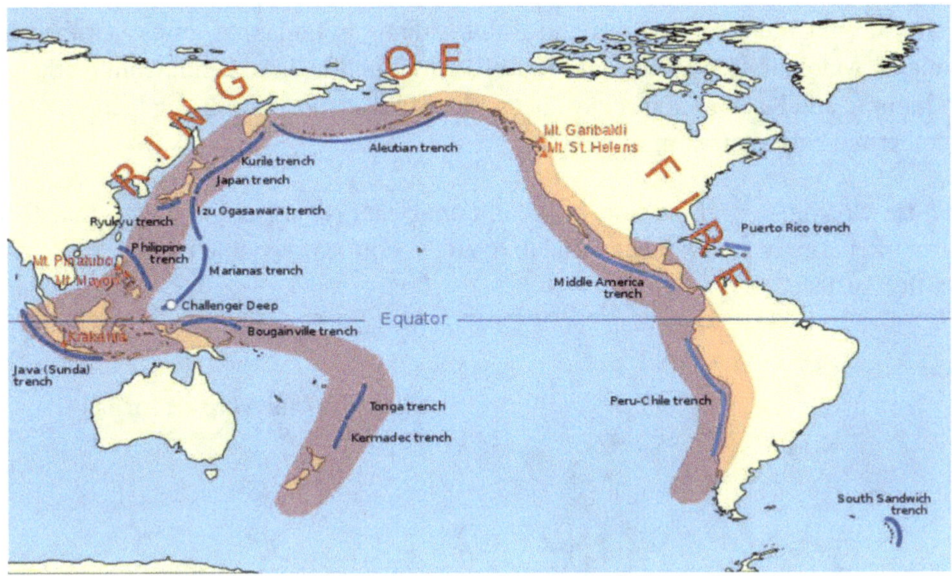

Fig. 9.37: Location of Most Active Geothermal Sites [107]

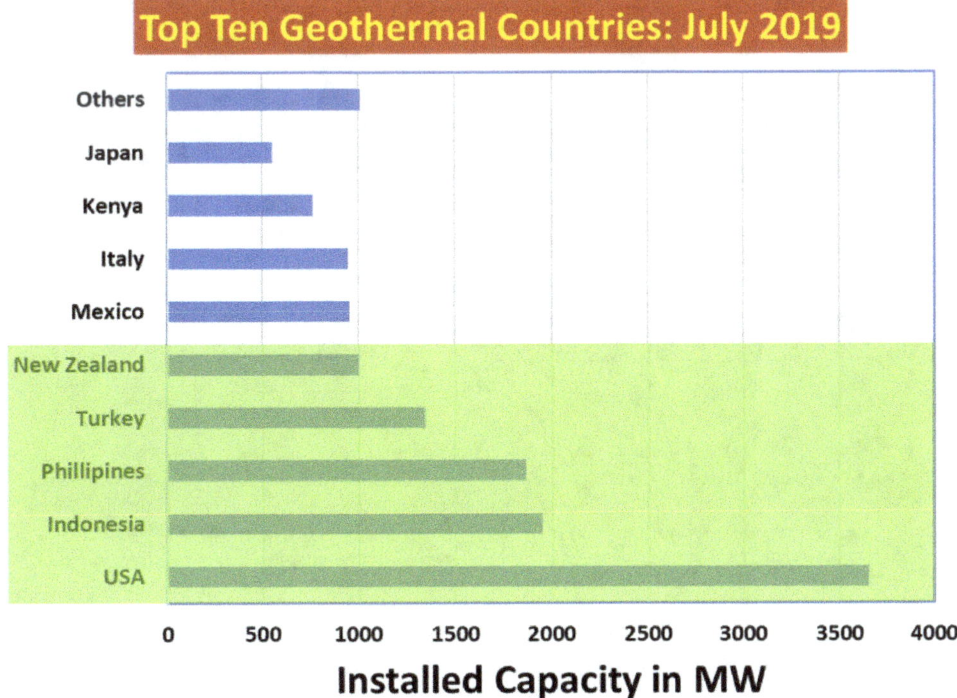

Fig. 9.38: Global Installed Capacity of Geothermal Energy, July 2019 [108]

Tattapani in Chhattisgarh, Puga and Chhumathang in Ladakh, Cambay Graben in Gujarat, Manikaran in Himachal Pradesh, Surajkund in Jharkhand, Chhumathang in Jammu, and Kashmir areas have been identified as most promising locations for geothermal exploitation in India.

Before closing this discussion on geothermal energy, we reemphasize that this form of energy is not widely available in an economically viable manner closer to Earth's surface.

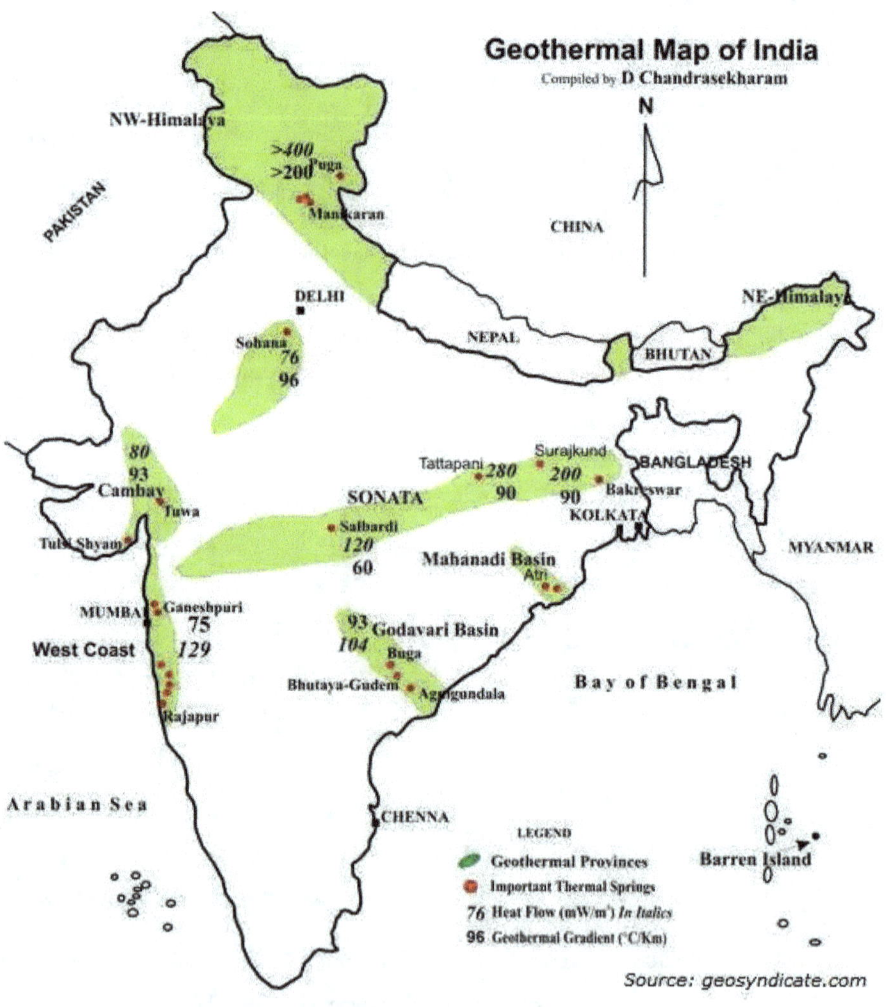

Fig. 9.39: Geothermal Map of India [109]

Thus, for example the success of Iceland in tapping the geothermal energy stems from the fact that Iceland is one of the most dynamic volcanic regions in the world. It was shaped by the activity of divergent tectonic plates which lifted heat and magma closer to the Earth's surface.

9.5 Tidal Energy

We know that we mostly have two high tides and two low tides in the seas, every day. At some places there is one high tide and one low tide. These are caused as the Earth goes around the Sun and the Moon goes around the Earth. Their gravitation pull does not affect the continents but pulls the water of the oceans towards them and thus the sea level rises, while decreasing in the perpendicular direction. As the Moon is nearer to Earth than Sun, its pull is about 2.2 times larger than that of the Sun — even though the Sun is about 27 million times more massive than the Moon. The high tides are highest when the Sun and the Moon are aligned as on New Moon (same side) or Full Moon (opposite side) day. These are lowest when their pull is perpendicular to each other.

Fig. 9.40: High Tide and Low Tide [110]

The difference in the sea level between low tide and high tide is largest near the coast and can vary from a few meters to about 15 metres. One can predict their timings and heights with a great deal of accuracy for months or years in advance. The periodic changes of water level cause tidal currents. One can also use these currents to generate power.

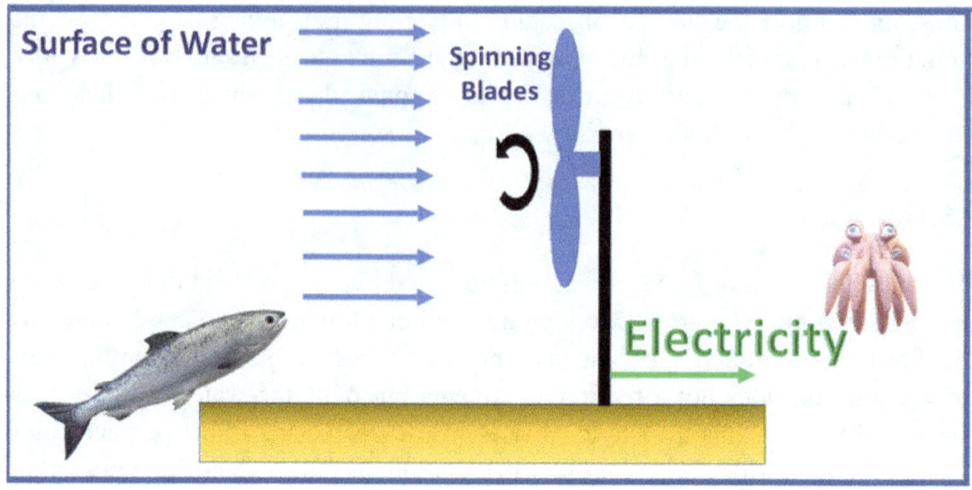

Fig. 9.41: Tidal Turbine [111]

Thus, tidal current energy is like wind energy, except that the density of water is about 800 times that of air and therefore a current moving at the same speed as air will produce 800 times more power than a windmill. Two methods are generally used for this. In the first, we put a turbine submerged in the sea water which is rotated by the current to produce electricity. One such location is Strait of Gibraltar, where the currents are quite strong and can be used to produce plentiful electricity.

These turbines can also be put at places where water flow is not due to tides, e.g., at the mouth of rivers at places where a river passes through a narrow opening between two hills or even sewage pipes in cities.

These water turbines are an interesting option as they do not generate noise like windmills and being submerged, they do not become eyesores. However, these can harm marine animals.

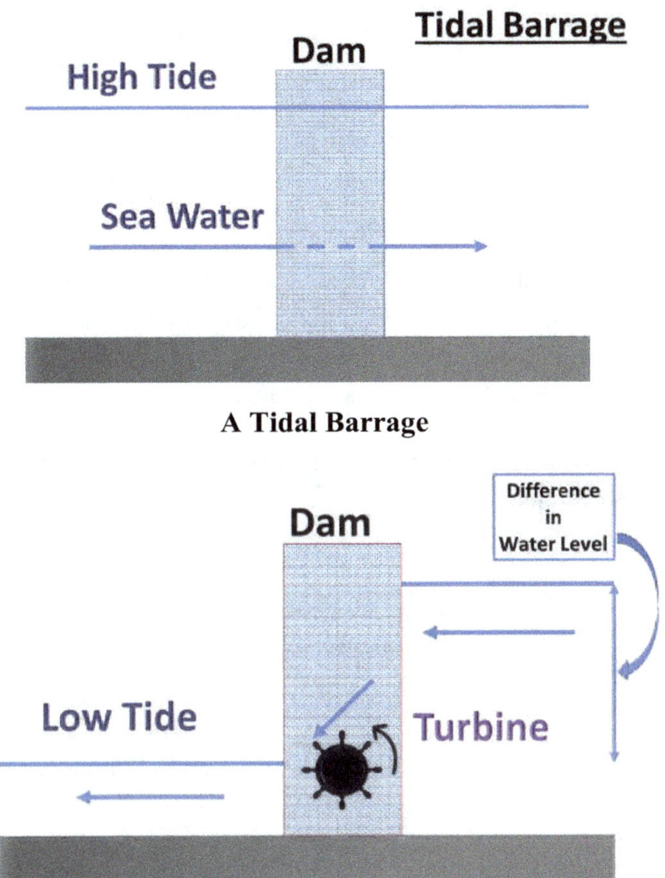

A Tidal Barrage

Fig. 9.42: Generation of Electricity using a Tidal Barrage [111]

The second option is tidal barrage. For this one makes a barrage with an opening so that the area behind the barrage is filled with sea water during high tide. Once full, the gates of the barrage are closed. The low tide then generates a level difference in the water on the two sides which is used to generate electricity. Tidal energy is a renewable source of low cast and clean energy. It does not use any fuel and has no waste by-products.

It was mentioned earlier that we know the timings of low and high tides in advance for months and thus unlike solar energy or wind energy which are unpredictable, tidal energy is highly predictable. If the turbines are used in a current of a river, they produce power continuously.

Additionally, tidal barrages can provide protection against flooding and land damage.

On the other hand, tidal barrages are needed to be installed at a site where the tides are consistently strong, to be economical. Even though quite predictable, the power generation is limited to only those times when the tides ebb and flow.

The barrage can accumulate silt, sediments, and pollutants within the tidal barrage from rivers and streams which are not able to flow out into the sea with low tides. We have already seen that marine animals can be hurt by the blades of the water turbine. They could also get stuck in the barrage.

We also need to develop turbines which can survive for long in the hostile environment of sea water.

India has a long coastline of about 7,500 Kms. It is estimated that the country can produce 7,000 MW of power in the Gulf of Khambhat in Gujarat, 1,200 MW of power in the Gulf of Kutch in Gujarat and about 100 MW of power in the Gangetic delta of Sunderbans in West Bengal.

The estimates of the global potential for tidal energy range from 150,000 MW to 300,000 MWs.

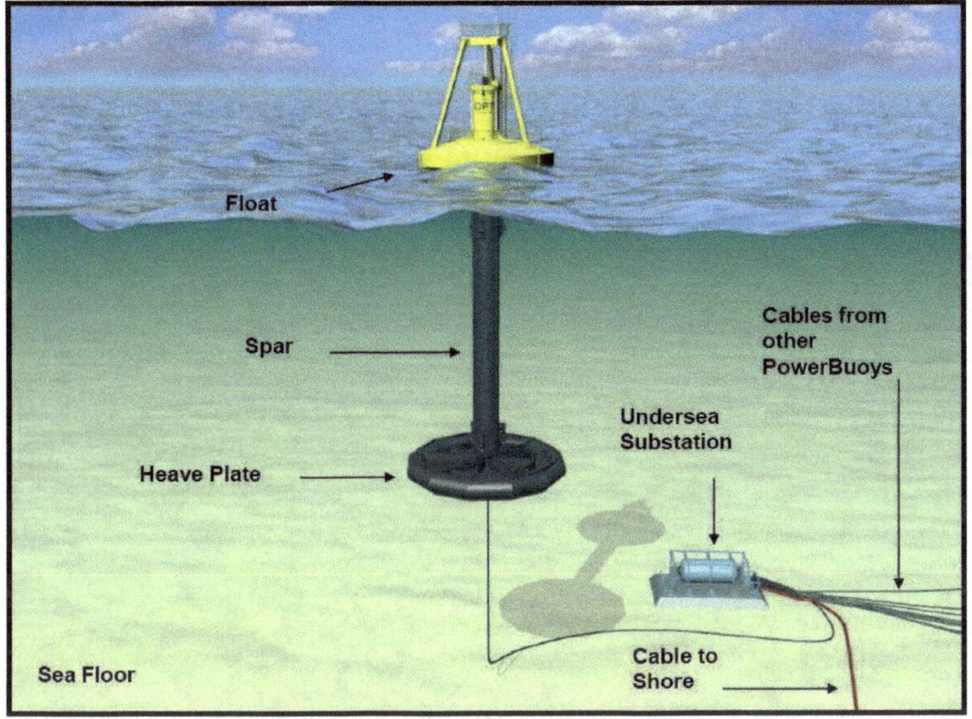

Fig. 9.43: Wave Energy [112]

A related idea involves utilizing the constantly rising and falling waves in oceans, where the up and down movement of a float or buoy is converted to electricity. We know that waves are always formed on the oceans, be it night or day. Even though, in principle, wave energy has an enormous potential — as 70% of the surface of the Earth is covered with oceans, only a few installations have come up so far, which use variations of the above arrangement, due to its interference with the environment, marine life, and even the pristine look of oceans.

9.6 Hydroelectric Energy

Fig. 9.44: Asteroids Brought Water to Earth [113]

Earth is the only planet with liquid water in our solar system, which has made life possible. Much of this water is believed to have been brought to it by comets, asteroids and planetesimals which rained upon Earth during its early history. The temperature on the Earth and its gravity have stopped this water from escaping to Space by evaporation.

This water brought to us by asteroids and planetesimals, filled up the oceans, formed lakes, got deposited into subterranean aquifers, evaporated, formed clouds, and came down to the Earth as rains and snow, and formed rivers and snow-capped mountains. The evaporation from the water bodies and oceans and transpiration from vegetation sends the water back to clouds and the cycle has been continuing. As the rain of asteroids stopped a long time ago, we do need to ensure that we use this water with care. (Some recent work suggests that the Earth may have had more water trapped in its rocks than previously assumed.)

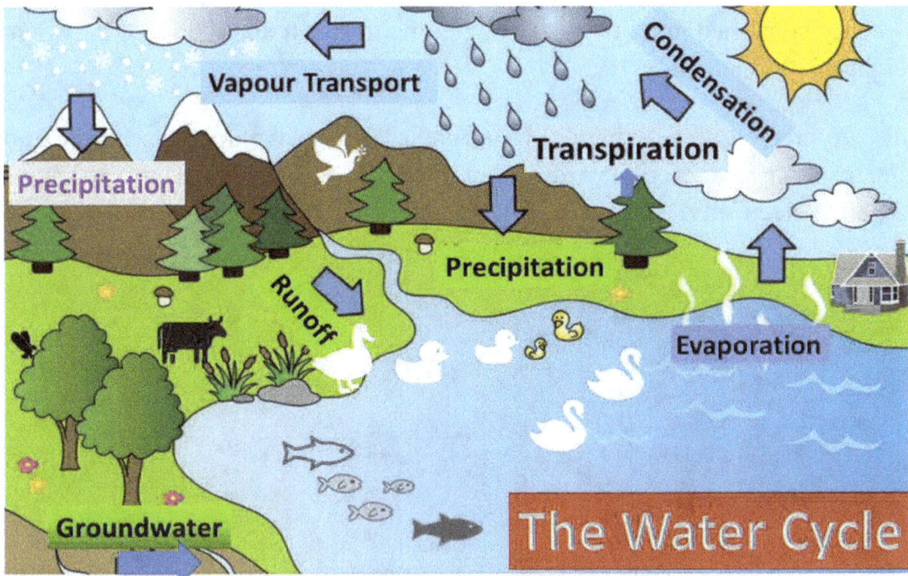

Fig. 9.45: The Water Cycle [114]

The total amount of water on Earth is estimated to be about 1.38 billion km^3. Out of this 97.5% (or 1.35 billion km^3) is in the oceans and just 2.5% (or 35 million km^3) is fresh water. About 1.7% of the total water on Earth (about 24.4 million km^3) is locked in glaciers and just 0.8% (or 10.6 million km^3) is accessible to us. This water circulates through the well-known water cycle.

Let us see, how this water is used to generate electricity. As the name suggests, hydroelectric power is generated by running water or water falling from a height to generate electricity.

A mass of water m falling from a height h would provide energy equal to $m.g.h$ where g is acceleration due to gravity (about 10 meter/second/second). Consider one cubic metre of water. Its mass is about 1000 kg. Let us assume that the height of the dam is 100 metres. Thus, it will pack (1000 kg) × (10 m/s2) × (100 metre) or 10^6 Joules (or 1 MJ) of energy. Therefore, a 100-metre-high dam releasing 1 cubic metre (1000 litres) of water per second will produce 1 MJ/sec or 1 MW of power. To generate 1 GW of power, its flow rate must be 1000 cubic metres/second.

These calculations also bring out the main problem with the gravitational energy — it is incredibly weak compared to chemical energy. For example, to get the amount of energy stored in a single AA battery (about 12,960 Joules), one will need to lift 100 kg of water to a height of 10 metres. One can show that to match the energy contained in a litre of gasoline, one must lift about 35 tons of water to a height of 100 metres!

Dams are erected across rivers, often, when these pass through a mountainous range, resulting in a reservoir, from where water is released in a controlled manner to generate electricity.

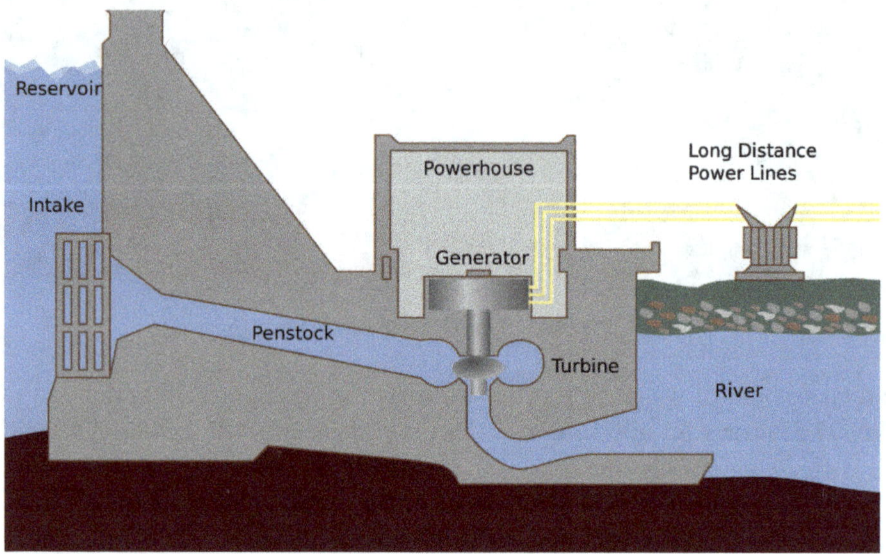

Fig. 9.46: Principle of Hydroelectric Dam [115]

These reservoirs could also be created using pumped-up water, taking power from a solar power plant or a windmill farm at a time when there is extra power and using it at a more convenient time to produce electricity.

One can have hydroelectric dams having capacities varying from just a few tens of kilowatts to provide power to isolated hamlets to those producing several thousands of megawatts of power.

Hydroelectricity is a reliable and renewable source of energy. Carbon dioxide is produced only during the construction. However, if vegetation is not cleared before the dams flood the area, it slowly rots and continues to release methane. The oxidation of the vegetation uses up the oxygen dissolved in water and affects the marine life.

Studies have also revealed that there can be a considerable loss of water due to evaporation — especially during hot months and in arid regions. It has been estimated that this loss could be 5000 litres to 70,000 litres/MWh. Evaporation leaves the salts dissolved in the water behind and the remaining water becomes more saline.

Of course, the water supply to the dams is affected by uncertainties in rainfalls and droughts.

Several of these issues can be addressed by using the so-called "float photovoltaics" or "aqua photovoltaics", where floating solar power farms are set up on the dams. It is estimated that this can reduce the evaporation loss by up to 70%, keep the temperatures of the solar panels low — thus increasing their efficiency, and make much more water available for production of electricity as well as for irrigation. These do not require extra land for installation of the panels and the water necessary for washing the panels is not wasted. The dust level over dams or large bodies of water are necessarily low. The covering of the water by the panels is also expected to reduce formation of algae, though its effect on marine animals is not yet sufficiently studied. Some countries have also considered putting such solar farms in the open seas, which, however, requires more permanent structures.

The resistance to large hydroelectric dams comes from the large areas which get inundated when lakes are formed. This destroys biologically rich and often productive farmlands, or forests, along with their ecosystem. It may also result in large scale displacement of population. Tehri Dam in Uttarakhand and Nagarjuna Sagar Dam in Andhra Pradesh drowned historically important cities and monuments, some of which were saved by shifting.

The World Commission on Dams estimated that dams had physically displaced 40–80 million people worldwide by the year 2000.

Dams disrupt the flow of water in the rivers across which they are built and affect the movement of marine animals. Quite a few varieties of fish, e.g., salmon and hilsa, migrate from the sea up rivers, where they spawn and then migrate back to sea. Construction of dams impedes their movement.

The disruption of the flow also causes deposition of silt in the dams reducing their capacity.

Fig. 9.47: Floating Solar Power Station in Napa Valley, California [116]

Care must be taken while choosing the site for making a dam. Tehri Dam, for example is close to a geological fault line, and close to the epicentre of an earthquake of the magnitude of 6.8 on Richter scale. Technical teams have confirmed that it is safe against earthquakes of a magnitude of 8.4 on the Richter Scale. However, some seismologists have suggested that earthquakes of the magnitude 8.5 can take place in that area. If such a catastrophe were to take place, the dam-break would submerge numerous towns down-stream, risking about half a million people. Dam breaks could also happen, if the water pressure is too high or if there is any fault in construction.

In 1975, during a typhoon the Banqiao Dam in Southern China failed, as about a year's worth of rain fell within 24 hours. This resulted in a flood which killed 26,000 people immediately and 145,000 from epidemics which followed. Millions were left homeless.

For the first time in the history of Kerala, 35 out of 54 dams were opened in August 2018, following very heavy rains, which led to a severe flooding. It led to evacuation of about a million people and death of about 500 persons.

It has been estimated that only 30% of the hydrodynamic potential of Europe, 25% of North America, 20% in South America, 5% in Africa, 5% in the Middle East and 20% of Asia-Pacific has been utilized so far.

Norway produces more than 99% of its electricity with hydropower. New Zealand uses hydropower for 75% of its electricity.

India's economically exploitable and viable hydroelectric potential is estimated to be 148,700 MW. It is also estimated that smaller hydroelectric power station of 25 MW or lower capacities can provide additional 6,789 MWs of power. 56 sites for pumped storage schemes with an aggregate installed capacity of 94,000 MW have also been identified.

Table 9.2: Top Five Hydroelectric Power Stations in the World

Rank	Power Station	Country	Capacity (MW)
1	Three Gorges	China	22,500
2	Itaipu	Brazil/Paraguay	14,000
3	Xiloudu	China	13,680
4	Guri	Venezuela	10,233
5	Belo Monte	Brazil	10,011

The Three Gorges power station is the biggest power station of any type in the world. It covers an area of about 1,045 square kilometres and required relocation of 1.24 million people. It is estimated that at full power, Three Gorges reduces coal consumption by 31 million tons per year, avoiding 100 million tons of greenhouse gas emissions. It also avoids millions of tons of dust, one million tons of sulphur dioxide, 370,000 tons of nitric oxide, 10,000 tons of carbon monoxide, and quite some amount of mercury. Hydropower saves the energy needed to mine, wash, and transport the coal from mines.

Some of the main hydroelectric power plants in India are also given below (Table 9.3).

Table 9.3: Main Hydroelectric Power Stations of India

Name	Location	Capacity (MW)
Tehri Dam	Uttarakhand	2,400
Koyna	Maharashtra	1,960
Srisailam	Andhra Pradesh	1,670
Nathpa Jhakri	Himachal Pradesh	1,500
Sardar Sarovar Dam	Gujarat	1,450
Bhakra Nangal Dam	Himachal Pradesh	1,325

Fig. 9.48: Sardar Sarovar Dam, Gujarat [117]

The utility of dams should not be decided by the production of electricity alone. These provide drinking water as well as water for irrigation to vast areas. The Bhakra Nangal Dam and the Nagarjuna Sagar Dams, *etc*., played a very important role in ushering in the Green Revolution in India.

These were famously described by the first Prime Minister of India, Pundit Jawahar Lal Nehru as “Temples of Modern India”.

Fig. 9.49: Canal Top Solar Panels for Narmada Canal Net Work [118]

The Sardar Sarovar Dam project additionally adopted an interesting innovation to take advantage of the sprawling canal network by covering it with solar power panels, thus converting adversity (of loss of water by evaporation) to an advantage.

An entirely new and important role for hydroelectric dams is now opening with the large-scale use of intermittent renewable sources of power like solar power or wind power. The water stored in these dams can be released for electricity production to coincide with the dip in production from solar and wind power plants but rise in demand for electricity, much like the pumped-up hydroelectric power systems, without seriously compromising their role as provider of water for irrigation. Thus, it can prove invaluable for balancing wind and solar power fluctuations because it can respond quickly and provide electricity as needed.

Chapter 10

Solar Energy

Sun has always given us light and warmth since the beginning of life on Earth. We have already seen that Sun is also the basis of all our energy resources — the hydrocarbons, water, wind, tides, waves, biomass, *etc.*

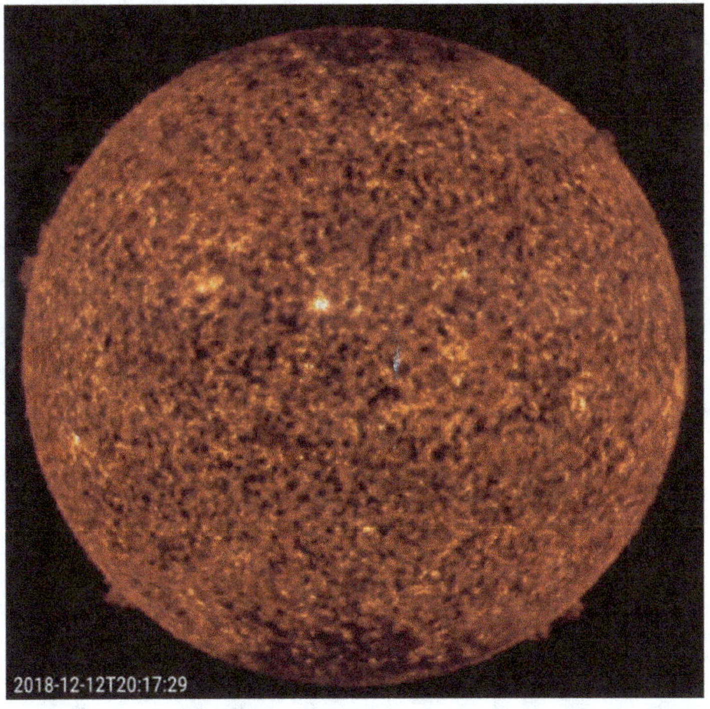

Fig. 10.1: Sun [119]

It is not a surprise that Sun is revered in all civilizations. Rigveda declares:

"सूर्य आत्मा जगतस्तथुषश्च"

(*Sun is the soul of the World.*)

Solar energy is one of the most hotly discussed and enthusiastically researched renewable energy in recent times.

The Solar Energy Potential

Our Sun is a medium sized star, with a diameter of approximately 1.391 million kilometres. It is a giant nuclear fusion reactor, converting about 600 million tons of hydrogen into 595.72 million tons of helium per second and releasing about 3.85×10^{26} Watts of power. The temperature of the Sun is estimated to be about 15 million degree Celsius at the core and about 6,000 degree Celsius at the surface. It is estimated that it will take 5 billion years before the hydrogen in it is exhausted, and helium burning starts converting it into a Red Giant, when the Sun will start swallowing all the inner planets including the Earth. We survive on Earth's surface since we are about 150 million kilometres away from the Sun and the intensity of solar radiation on its surface is only about 1362 Watts per square metre. Considering that the Earth reflects about 30% of this radiation back into space, the Earth's surface receives only about 1000 Watts per square metre of solar energy.

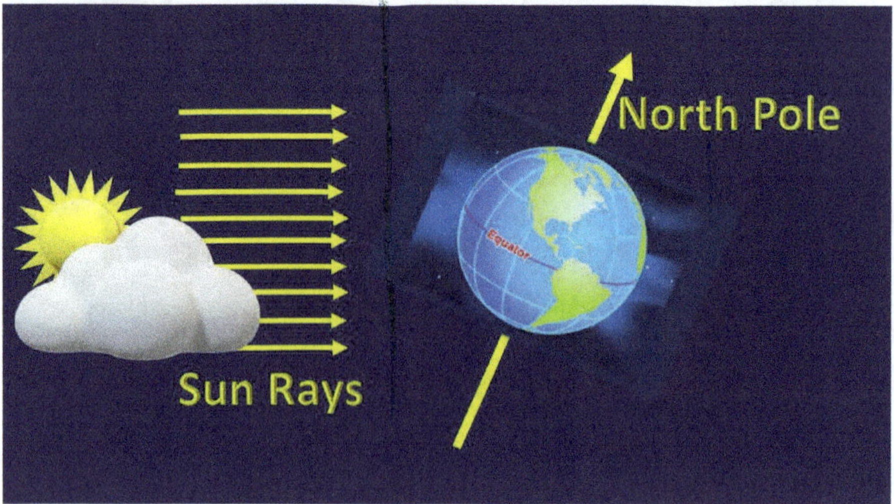

Fig. 10.2: Sun Rays Strike the Earth's Surface at an Angle [120]

However, this is for vertical incidence of the Sun's ray at the Equator and we need to take an average over the angle of incidence as one goes from the Equator to the Poles. Further, one must take the seasonal variations due to the tilted axis of the Earth into account.

In addition, we need to consider the spectrum of the solar radiation and its absorption by the water vapour and other gases in the atmosphere. The radiation from the Sun closely corresponds to a black body radiation at 5250 degree Celsius. We will see this if we measure it at the top of the atmosphere. The sunlight reaches us after traversing the large body of air. Consider midday Sun at the Equator, when the sunlight would be perpendicular to the surface of the Earth. This body of mass of air is designated as AM1.0 (air mass, 1.0). At later or earlier times or at places away from the Equator, this sunlight will strike the Earth at an angle and will necessarily cross a larger path length. We see that the atmosphere removes much of the ultra-violet end of the spectrum and saves us from their harmful effects with the help of ozone layer discussed earlier. Several bands in the visible part of the spectrum are also absorbed by the water vapour in the atmosphere.

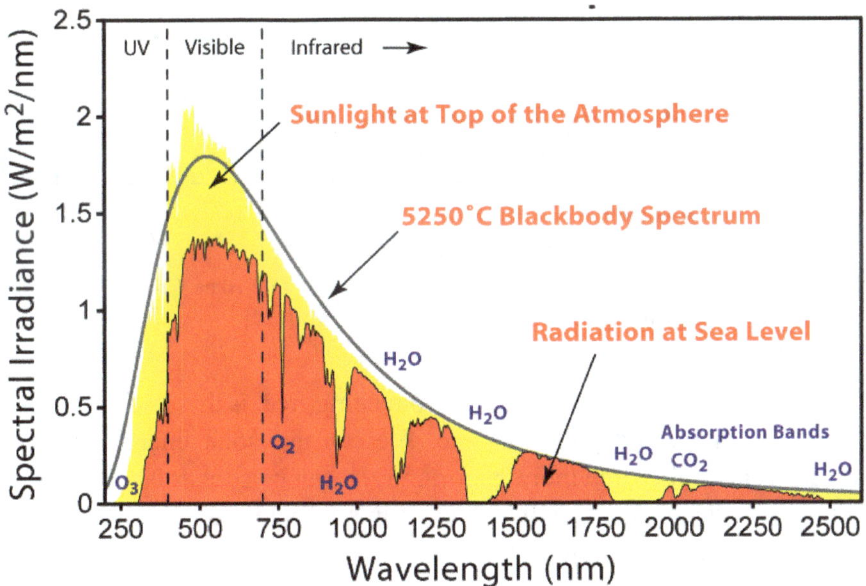

Fig. 10.3: Solar Radiation Spectrum at Sea Level After Going Through 1 Air Mass (AM1.0) [121]

Taking all the above effects into account, one can calculate that most of the World population lives in insolation areas of 150–300 W/m^2, which corresponds to an energy of 3.5–7.0 kWh/m^2 per day.

One can easily calculate the energy received by the Earth by multiplying the above with its surface area. Considering that the radius of the Earth is about 6,400 kilometres, the energy received by the Earth is about 1.8 to 3.6 million Tera Watt Hours/day. These estimates suggest that the solar energy received by the Earth in an hour is more than enough to meet the energy requirement of the entire world for a year.

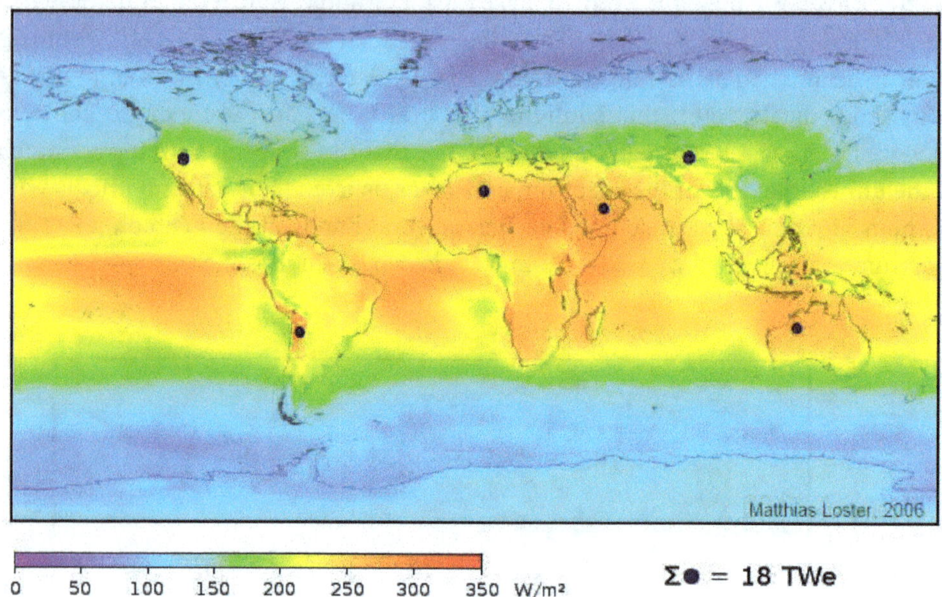

Fig. 10.4: Distribution of Solar Radiation across the World. The Area Covered by the Black Dots can meet the Energy Requirement of the Entire World! [122]

The greatest virtue of solar energy, apart from being free, is that it is inexhaustible, universally available, and pollution-free. We can use solar energy directly for heating (solar thermal). Even more importantly, it can be converted into electricity using the heat to generate steam which can run a turbine. Solar photovoltaic cells can be used to generate electricity directly.

10.1 Solar Thermal

Olympic Games, which started in 776 BC, used to have an important ritual. Eleven girls, representing the Vestal Virgins, used to light the Olympic Torch kindled by the light of the Sun concentrated by a parabolic mirror at the Temple of Hera in Ancient Olympia. The tradition is still followed.

Fig. 10.5: Lighting of Olympic Torch by a Parabolic Mirror [123]

Legend also has it that Archimedes used parabolic reflectors to burn Roman ships which came to attack Syracuse during the Second Punic War in 212 BC.

Even though some modern historians have doubted the accuracy of this episode, the legend has survived and has been behind many studies on concentration of heat of the sunrays since ancient times. In today's context, solar heat can be used directly as heat or used to produce steam and run turbines to produce electricity.

Solar thermal systems can be either active or passive depending on whether they involve moving mechanical parts or not. Passive systems, like greenhouses, use specific design features to capture heat. These also include special designs for buildings which reduce the energy required for heating or cooling.

Fig. 10.6: Archimedes Burning Roman Ships Using a Parabolic Mirror, 212 BC [124]

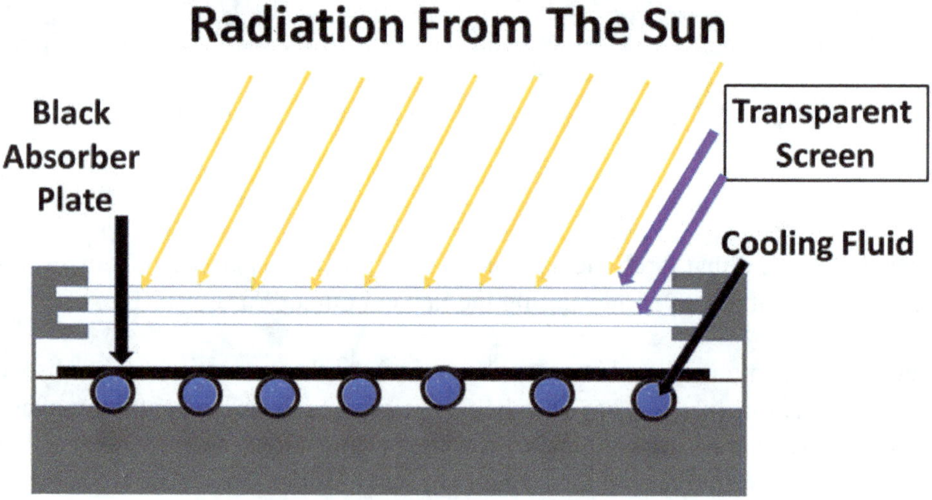

Radiation From The Sun

Black Absorber Plate

Transparent Screen

Cooling Fluid

Fig. 10.7: A Typical Flat Solar Collector [125]

Heating of water for domestic and industrial uses is one of the most popular applications of solar radiation. These systems are simple, easy to install, require very little maintenance, and use flat plate solar collectors. It consists of a flat plate to absorb radiation from the Sun. This is normally put under a transparent sheet of glass or plastic to reduce heat loss due to radiation or convection. Tubes are attached to the plate to transport the circulating water. These tubes are laid on an insulated bedding to reduce heat loss.

Solar water heating systems are used across the world extensively and may be saving as much as 17% of electricity use.

Solar energy can also be concentrated using flat plates, discs, or parabolic trough to get temperatures reaching up to 350 degree Celsius and can be used for a variety of similar applications such as cooking, drying and pasteurization. Tracking of the Sun as it moves across the sky helps in harvesting more heat during the day.

Fig. 10.8: A Solar Thermal Heater

All these applications reduce demand for fuel or firewood and help improve air quality by reducing a possible source of smoke.

An interesting concentrating technology known as solar bowl has been developed and deployed in Puducherry. The system uses a fixed spherical reflector with a receiver which tracks the focus of light as the Sun moves across the sky and can reach temperatures of 150 degree Celsius.

Fig. 10.9: Solar Bowl in Auroville, Puducherry [126]

Scheffler reflectors employ a parabolic dish with a single axis tracking to follow the Sun's daily course. Their reflective surfaces change their curvatures to adjust to seasonal variations of the incident sunlight. They have the additional advantage of having a fixed focal point which improves the ease of cooking and can reach temperatures of 450–650 degree Celsius.

Higher temperatures are also valuable for producing electricity using turbines. This is achieved in a power tower by using heliostat mirrors which are usually plane mirrors, which are tipped up and down and rotated from east to west to keep reflecting sunlight toward a fixed point, by compensating for the Sun's movement across the sky.

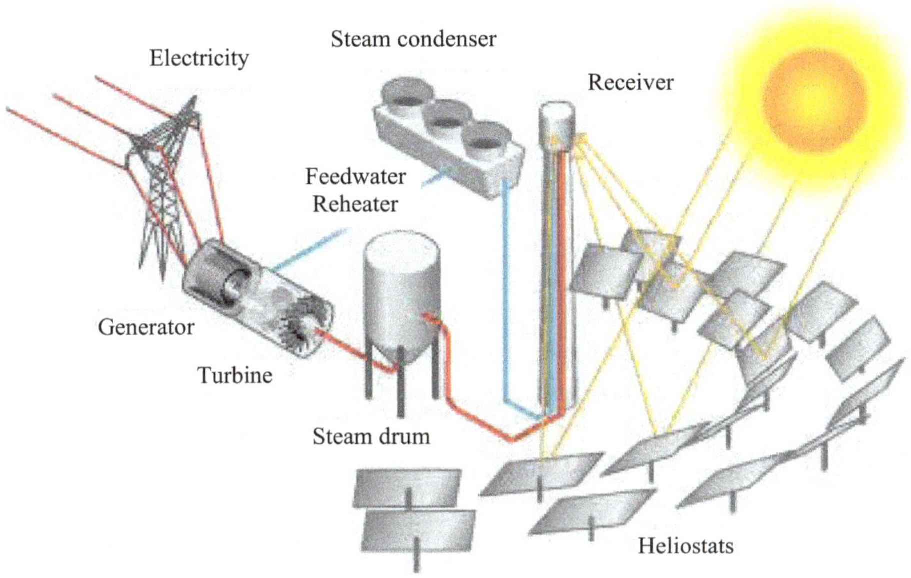

Fig. 10.10: Principle of Power Towers [127]

Fig. 10.11: Ivanpah Solar Power Tower Facility [128]

For example, the Ivanpah solar power tower facility in Mojave Desert in California has been producing up to 390 MWs of power (annual output 798 GWh) for several years now. It has three towers and a total of 173,500 heliostat mirrors. The concentrated sunlight is used to heat molten salt (a mixture of sodium nitrate, potassium nitrate, and calcium nitrate) to near 540 degree Celsius and to generate steam. The molten salts can also be stored in insulated tanks to be used for production of electricity on demand for up to thirty days. Thus, these can be thought of as batteries (of heat) to be used at any time.

These plants need to be stationed in remote and uninhabited area, as these do require huge areas (the Ivanpah station has an area of 3,500 acres). In deserts or arid regions, this raises questions of availability of water, both for heating as well as for cleaning the surface of the mirrors.

It is interesting to note that the concept of solar furnaces have also been used to concentrate power of "10,000 Suns" and reach temperatures of the order of 3500 degree Celsius, for the study of very special materials at very high temperatures.

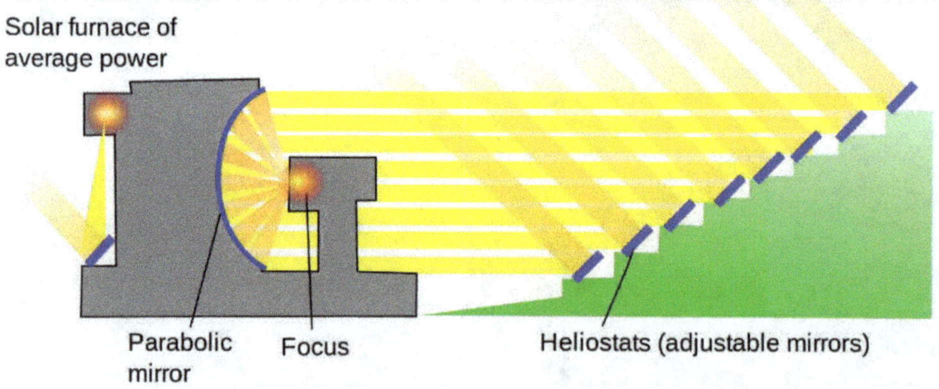

Fig. 10.12: Solar Furnace for 1 MW of Power [129]

Solar furnaces of even smaller powers have been operating across the globe in several locations. One such solar furnace has been operating at Odeillo in Pyrenees-Orientales in France for several decades.

Fig. 10.13: The Solar Furnace at Odeillo in the Pyrenees-Orientales in France [130]

Another concentration strategy for solar thermal power stations is to use parabolic troughs to focus sunlight along a tube running through the focal points of the reflectors.

Parabolic trough reflectors are made by bending a sheet of highly polished reflective material to a parabolic shape. The rays from the Sun are essentially parallel to each other when they hit the Earth. The parabolic shape of the reflectors focusses these along a line where one puts a heat absorbing pipe containing oil. This pipe can be a Dewar tube. It can contain water, oil, or as in modern technology available nowadays — molten salt, for increased efficiency. The Dewar tube provides that there is no heat loss due to conduction or convection. In the absence of the Dewar tube, the first loss can be high as the temperature rises, while the second loss can be substantial if there is a breeze or a strong wind. For best results, these reflectors must face the Sun directly. This is achieved by connecting them to a Sun tracking device.

The solar trackers generally have a single rotation axis along the length of the trough which can be orientated in east-to-west direction, tracking the Sun from north to south, or orientated in north-to-south direction and tracking the Sun from east to west.

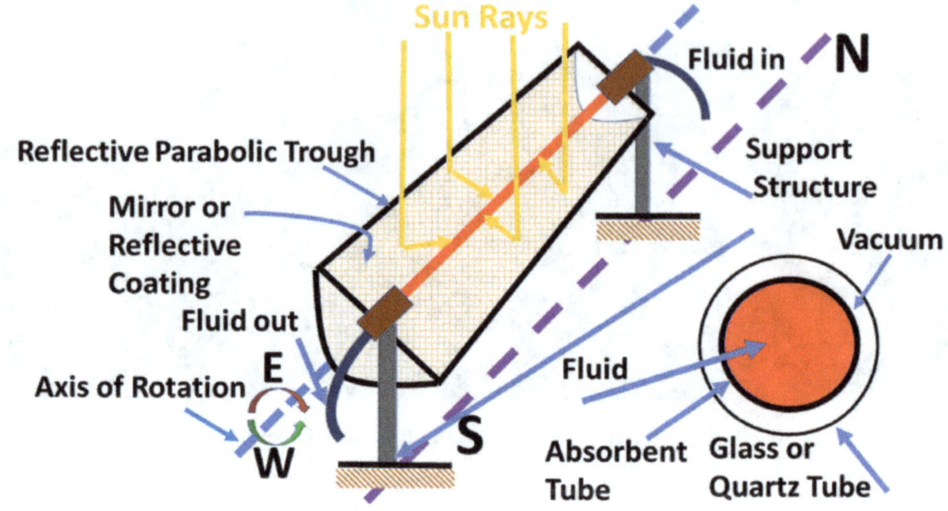

Fig. 10.14: Parabolic Trough Reflector [131]

Parabolic trough reflectors are modular, and one can easily cover a large area for the collection of the solar energy. These are connected in series and parallel for ease of heating and collection of the oil. The oil can reach temperatures of up to 400 degree Celsius and the molten salts still higher and can be used to produce steam at high temperatures.

Deserts tend to have frequent strong dust and windstorms. While strong dust storms cover the mirrors with dust needing immediate cleaning, windstorms may damage the mirrors. The 354 MW Solar Energy Generating Systems in California loses up to 3,000 mirrors every year, which need to be replaced. This is minimized by turning the mirrors during storms. The parabolic reflective panels also need to be cleaned regularly to maintain their high reflectivity.

Fig. 10.15: Parabolic Trough Solar Collectors at Solar Energy Generating Systems, California [132]

There have been reports of death of thousands of birds every year as they are incinerated in mid-flight when they enter the searing-hot solar fields. There have also been reports of aeroplane pilots getting disturbed by the glare. At least once some misaligned mirrors focussed the sunlight to a vulnerable point and caused a fire.

Parabolic trough solar collectors are among the cheapest options for collection of solar energy. By now a considerable experience has been accumulated in operating them. It is worthwhile to mention that one needs to be careful that the temperature of the molten salt, if used, does not fall below 120–220 degree Celsius (the exact value depends on the composition), below which it freezes causing considerable difficulty.

Fig. 10.16: Molten Salt Storage Tanks under Construction at Solana Generating Station, USA to Provide 280 MWs of Power for 6 hours [133]

Let us elaborate a little more on the molten salt storage of heat (Fig. 10.16). By now it is known that molten salts may lose only about 1 degree of heat a day. This helps us store and "fill-up" this thermal energy for months. It may be useful to design the size of solar thermal energy storage so that the energy can be used for months. It is possible to heat and cool the thermal energy daily for up to 30 years. After that, the tanks may be repaired for corrosion if any and used all over again for 30 years.

A scheme of storing this heat as molten scrap aluminium — which cools and solidifies and melts again when heated has been discussed in the literature.

Linear Fresnel Reflector Concentrators are one other simple option to heat oil to temperatures of about 250 degree Celsius and use it for production of steam/hot water for use in industry.

These are made from straight strips of reflecting material, which could be a metal coated with metallised plastic or mirror of glass, or a highly polished reflecting metal. These strips are oriented to focus the reflected light on a point (line), mimicking a concave mirror.

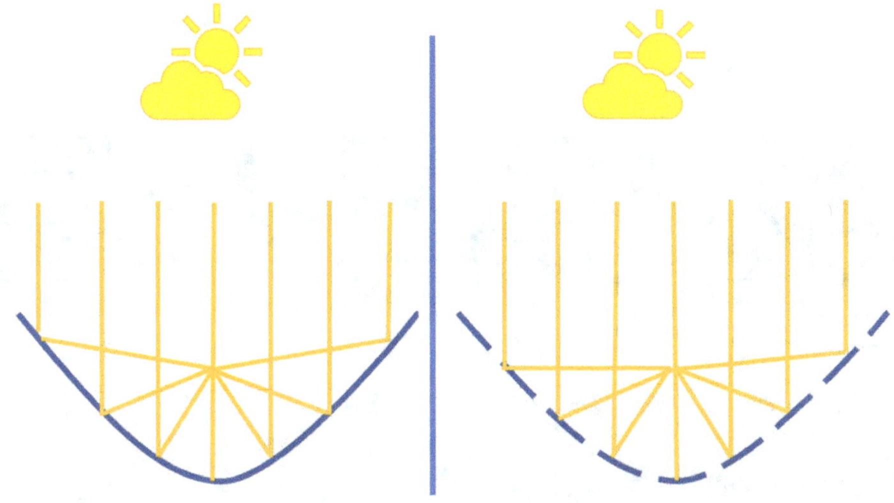

Fig. 10.17: Principle of Fresnel Linear Reflecting Concentrators [134]

These strips must be sturdy enough to be placed in the open and subjected to wind, rain, and heat, without loss of their reflectivity. As shown above, the strips of reflectors are oriented such that the reflected sunlight is collected by a metallic collector pipe which runs axially above the array of reflectors. These pipes may be coated with special materials to increase absorption of sunlight by them and may be of Dewar type as earlier to reduce the conductive and convective heat loss.

The orientation of the reflectors needs to be continuously adjusted so that reflection reaches the collector tubes. The collector tubes are then connected as in the case of parabolic trough collectors discussed above and their energy can similarly be stored as molten salts or even steam.

Fig. 10.18: 100 MW Linear Fresnel Reflector Concentrating Solar Power Plant, Rajasthan [135]

The largest Linear Fresnel Reflector Concentrating Solar Power Plant in the world at the time of its commissioning in 2015 was in Rajasthan, having a capacity of 100 MW. These plants produce power more cheaply as they employ simpler technology and are more efficient. The land requirement for them is also lower.

Fig. 10.19: Solar Dish Stirling Power System [136]

The collected heat can also be utilized to run a Stirling engine mounted either at the receiving point or connected to the point to utilize the heat using molten salt, steam, or oil. Depending upon the area covered, the dish can attain extremely high temperatures, which improves efficiency of the engine. Stirling engines are becoming popular for power conversion along with the concentrating dish.

Stirling solar thermal power generation technology has several advantages. It has high efficiency. It does not occupy too much land and is eminently suited for modular combination. It can be easily built in modules of 100 kW and scaled up to 100 MW, tailored to a large range of locations — from remote and isolated to industrial.

Stirling engines are simple in concept and easy to maintain. Three engine types with slightly different layouts are discussed in literature and industrial model using slightly different combinations of the hot and cold chambers — namely alpha, beta and gamma Stirling engines — each with its advantages and disadvantages are available.

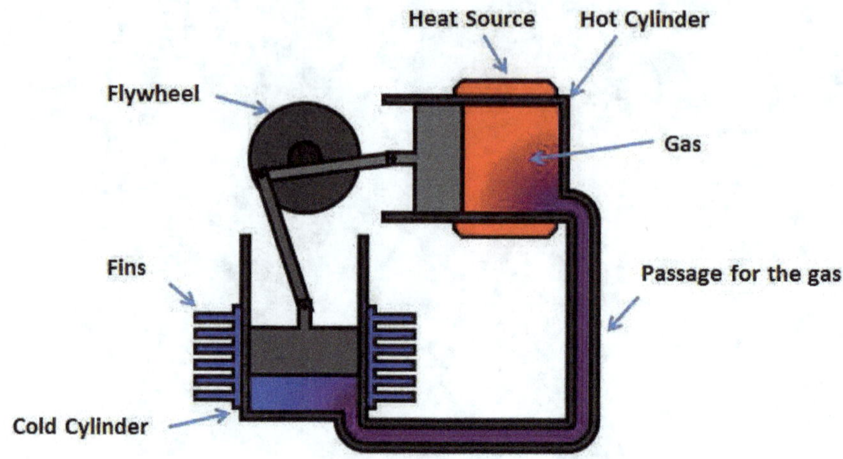

Fig. 10.20: Main Parts of Alpha Stirling Engine [137]

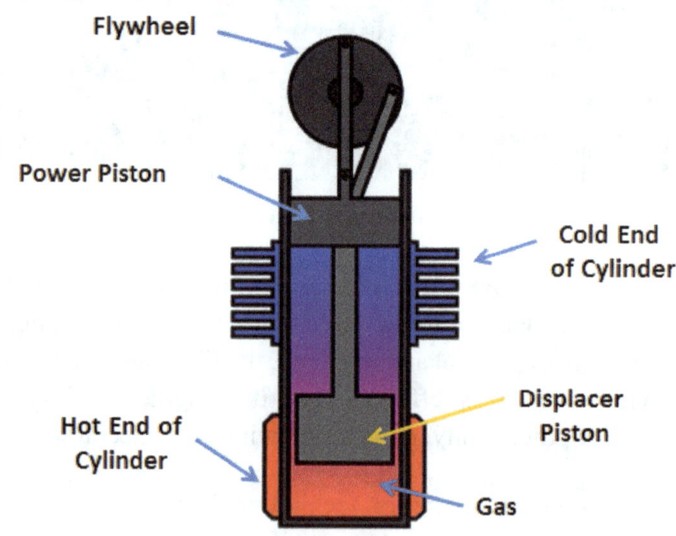

Fig. 10.21: Main Parts of Beta Stirling Engine [138]

In brief, Stirling engines work by heating and cooling of the gas in turn by an external heat source. It extracts energy from the gas's expansion and contraction. This causes change in pressure which operates a piston.

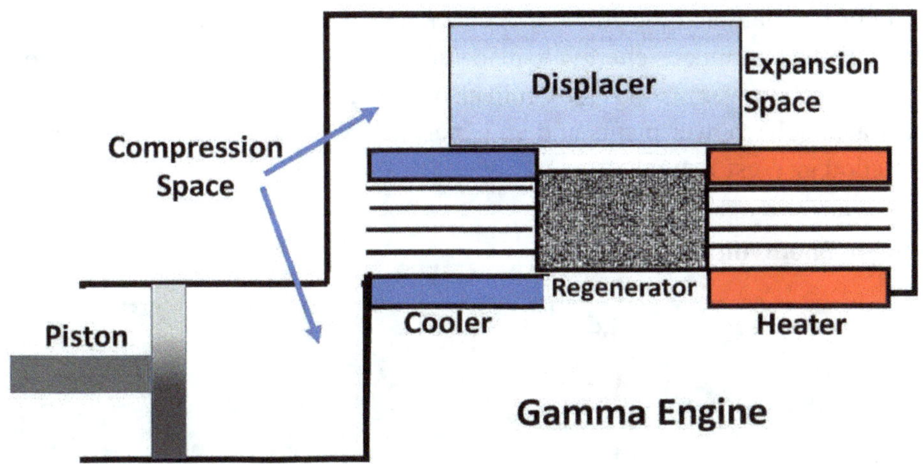

Fig. 10.22: Main Parts of Gamma Stirling Engine [139]

Before closing this discussion, we reiterate that concentrated solar power plants are a very green source of energy, which are quite efficient, and which can be built to provide any level of power from 10 kilowatts to hundreds of Mega Watts. These can be coupled with molten salt insulated containers which can store heat energy for up to 30 days, to be used for generation of electricity at any given time. By adjusting the overall capacity of the collector and the storage system, one can programme these to provide uninterrupted power.

We also reiterate that solar thermal plants utilize the entire spectrum of the Sun for heating. This conversion efficiency can be as high as 85–90%. Considering that up to 75% of the energy used by industries is used for heating, a direct utilization of the heat collected by solar thermal power plants can prove to be very valuable, efficient, and cost effective.

We further add that India is poised to decommission many very old, inefficient, highly polluting, and sub-critical coal fired thermal power stations. One could replace those with Concentrated Solar Plants. Initially they may use the existing turbine systems till those are replaced by more efficient modern turbines.

As mentioned above, along with molten salt insulated tanks and adjustment of capacity of the plant and the storage, one can have a very clean and green source of plentiful and continuous energy, in place of highly polluting and inefficient coal burning thermal power plants. One could also consider covering water bodies with concentrated solar power plants and an experimental pilot plant for this is under construction in West Bengal.

10.2 Solar Photovoltaics

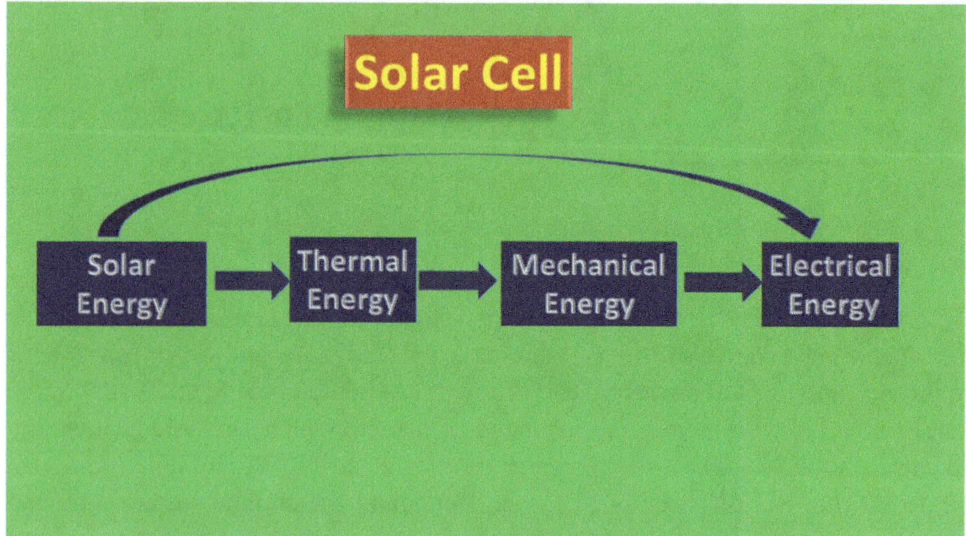

Fig. 10.23: Solar Cells Convert Solar Energy Directly to Electricity

Conversion of solar energy directly into electricity is very valuable since we lose energy whenever we change energy from one form to another. For example, electricity from coal requires burning of coal to generate heat, uses the heat to generate steam and then run a turbine, and finally the turbine produces electricity. Solar thermal systems need to generate steam, run a turbine using the steam and generate electricity. In both these cases, there is irrecoverable loss of energy at every stage. Solar photovoltaic cells convert solar energy to electrical energy directly using the well-known photoelectric effect or photovoltaic effect, known since early 1900's.

To understand photoelectric effect, we recall that using electrical conductivity one can classify materials as conductors, semiconductors, and insulators. Using the theory of band structure of materials, one can generally categorize electron energy levels into "valence band" and "conduction band".

The highest occupied electron energy levels populate the valence band, while the lowest unoccupied electron energy levels constitute the conduction band. The energy difference between the top of the valence band and the bottom of the conduction band is called "band gap" and its value serves as a very convenient parameter for categorizing materials as conductors, insulators, and semi-conductors.

In a conductor the valence and conduction bands overlap, and the band gap is zero. This allows free movement of electrons through the material as in metals like copper or gold.

In an insulator the band gap is quite large (of the order of 3 to 8 eV), making it very difficult for the electrons in the valence band to move to the conduction band, and facilitate conduction of electricity, as for example in wood, glass, and ceramics.

Semiconductors have a band gap of the order of 1 eV. This can be provided by sunlight which can move some of the electrons from the valence band to the conduction band, which can then facilitate conduction of electricity.

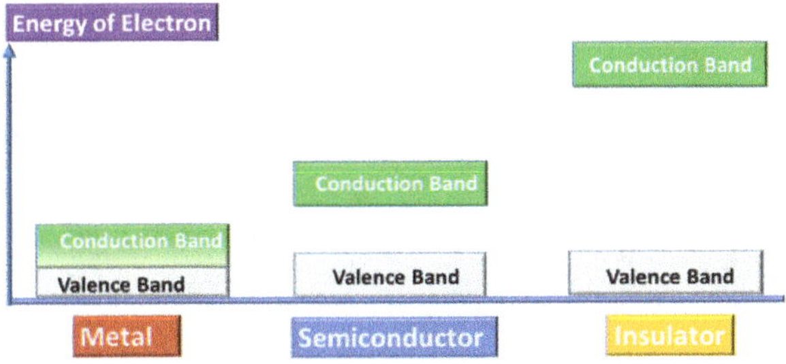

Fig. 10.24: Metals, Semiconductors, and Insulators [140]

Silicon and Germanium are two of the most studied semiconductors. We have already seen that silicon has a band gap of about 1.1 eV. Germanium has a band gap of about 0.7 eV. Germanium also has a better mobility for the electrons and holes, but it is expensive to produce and its crystal structure is easily disturbed at higher temperatures.

Silicon is very widely distributed as sand (silicon dioxide or silica). It makes up 26% of Earth's crust and is second most abundant element after oxygen. It is also the seventh most abundant element in the Universe. Its crystal structure is not easily destroyed.

The basic operation of solar cells can be understood as follows. As light falls on a semiconductor, some of it just passes through and some of it is reflected — while the remaining light is absorbed. If the energy of the photon is more than the band gap of the semiconductor, it knocks out an electron from the valence band, promoting it to the conduction band. A hole is created in its place which behaves like a positively charged particle. This electron-hole pair is called an exciton. The pair is broken by an internal or an external field and the charge is collected, which appears as an electrical current.

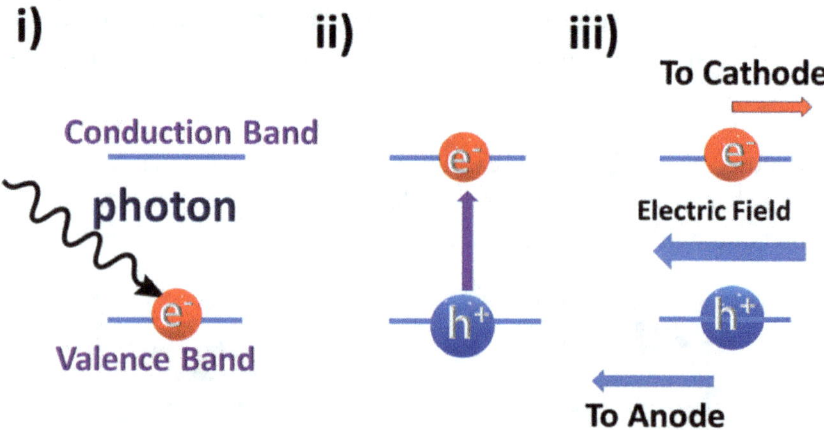

Fig. 10.25: Basic Principle of a Solar Cell [140]

In inorganic semiconductors such as silicon, the four outer electrons are shared with four other silicon atoms, forming covalent bonds. A small percentage of foreign atoms in the regular crystal lattice of silicon (or germanium) produces dramatic change in their properties, leading to what are called n-type and p-type semiconductors.

Replacement of silicon atoms with atoms having five valence electrons, like those of antimony, arsenic, and phosphorus gives rise to n-type semiconductors, as these have one extra electron per impurity atom. Replacement of silicon atoms with atoms having three valence electrons, like those of boron or gallium, gives rise to p-type semiconductors as these are one electron short per impurity atom. This process of replacing silicon (or germanium) atoms in the pure crystal is called doping. It is done either by diffusion or ion implantation.

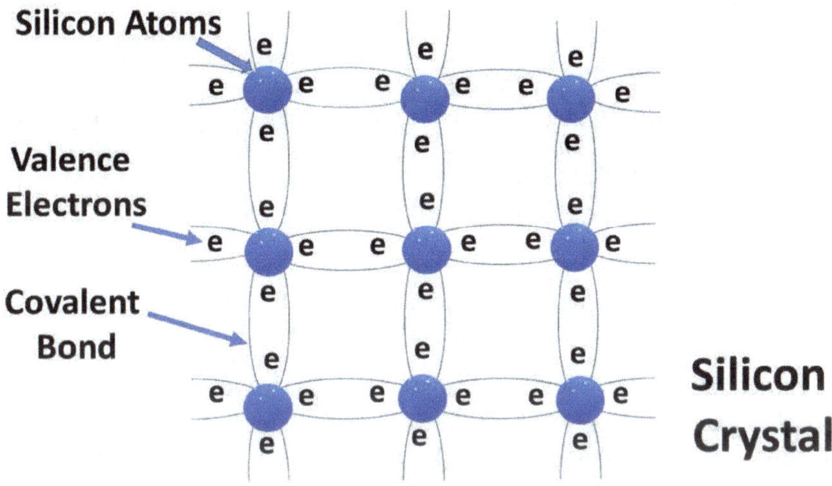

Fig. 10.26: Silicon [141]

When p-type and n-type semiconductors are in contact, charges build up on either side of the interface, as electrons move preferentially to the p-type semiconductor, and holes move to the n-type of the semiconductor. These semiconductors, by themselves have no net charge as the number of their electrons is balanced by the number of the protons in their atoms.

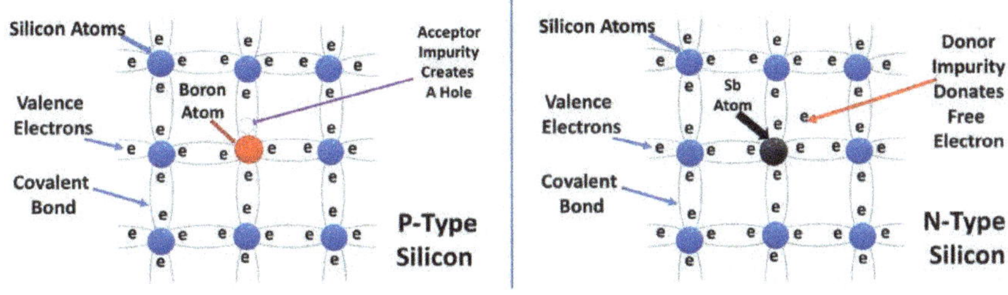

Fig. 10.27: A n-type (left) and a p-type (right) Impurity in Silicon [142]

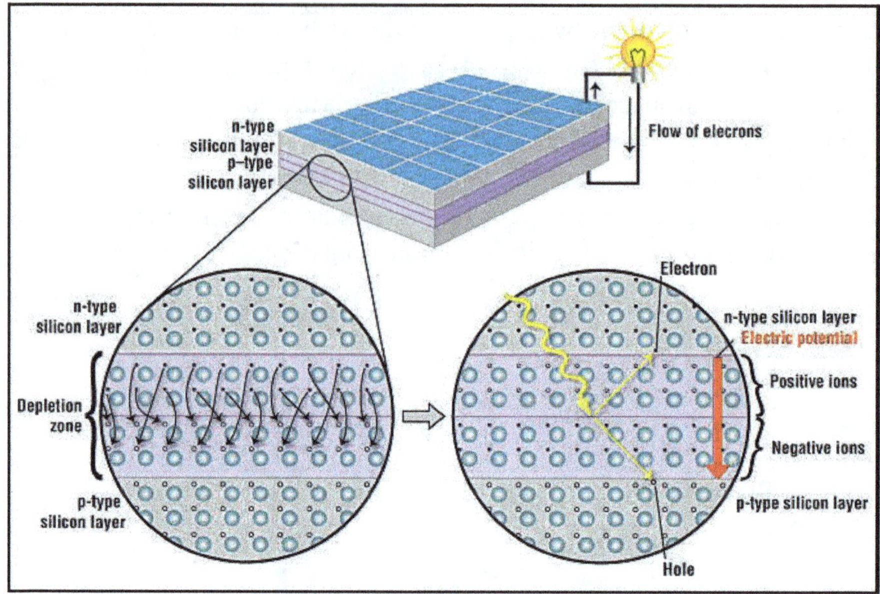

Fig. 10.28: Working of a Solar Cell [143]

This movement of electrons and holes completes the covalent bonds but leads to a build-up of negative charge on the side of the p-type semiconductor and a build-up of positive charge on the n-type of the semiconductor. This results in creation of an electric field pointing from the n-type to the p-type region, which now prevents electrons in the n-type layer to fill holes in the p-type layer.

The region of the contact is called depletion region or space charge region.

When sunlight strikes this cell, electrons in the silicon are ejected, which results in the formation of "holes" or vacancies left behind by the escaping electrons.

The field mentioned above moves the electrons to the n-type layer and holes to the p-type layer. If the n-type and p-type layers are now connected with a conducting wire, the electrons will travel from the n-type layer to the p-type layer by crossing the depletion zone and then go through the external wire back of the n-type layer, leading to a flow of electricity.

Thus, a typical solar cell will consist of a glass or plastic cover, an anti-reflective layer, a front contact to allow electrons to enter a circuit, a back contact to allow them to complete the circuit, and the semiconductor layers where the electrons begin and complete their journey.

The amount of power generated by a single solar cell is of the order of one to two watts. Therefore, solar cells are grouped together to supply a more useful level of voltage, current and power. The cells are connected in parallel to produce higher currents and in series to produce higher voltages.

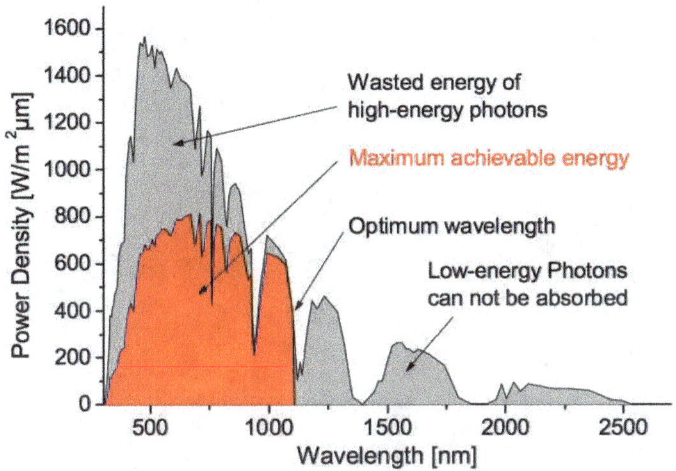

Fig. 10.29: Spectral Losses in a Solar Cell. The Figure shows the Maximum Achievable Energy of a Silicon Solar Cell in Relation to the Sun Spectrum (AM1.5) [144]

We have seen that crystalline silicon, the most popular semiconductor used for making solar cells has a bandgap of 1.1 electron volts. We realise that if the energy of the incident photon is more than this, the excess energy goes into heating the solar cell. Of course, if the energy of the photons is less than this, then it cannot kick the electron out. Thus, only a part of the solar spectrum is used for production of solar power. Further, as mentioned earlier as well, a part of the sunlight is reflected, and another part just passes though the cell.

Taking all these considerations into account William Shockley and Hans Queisser calculated the theoretical upper limit of efficiency of solar cells as a function of the band gap, which is called Shockley and Queisser limit.

Theoretically, the maximum possible efficiency of solar cells is now believed to be 33%. It has been estimated that it happens for a band gap of about 1.34 eV. It is also estimated that about 47% of the solar energy gets converted to heat, about 18% of the photons pass through the solar cell, and about 2% of the energy is lost to local recombination of the newly created holes and electrons.

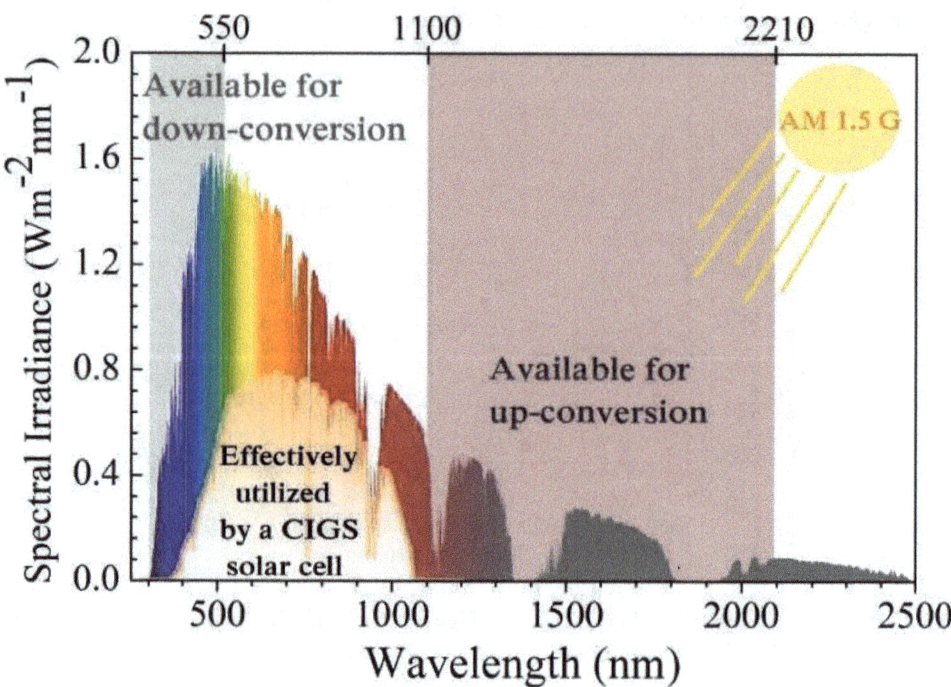

Fig. 10.30: Sunrays Absorbed by Copper Indium Gallium Selenide Solar Cells [145]

We have seen that photons having energies less than the band gap do not produce electrons and the photons having more energy than the band gap, lose part of the energy to heat. There have been suggestions to trap this heat. There have also been suggestions to apply a coating to the cells which can help convert two photons of lower energy to one with a higher energy. However, the expenditure for this can be rather high and efficiency is still low.

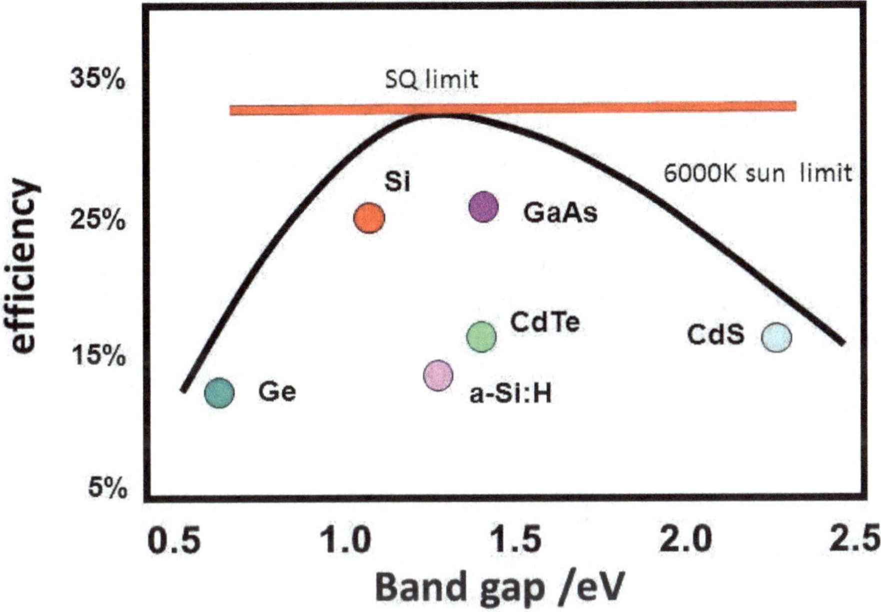

Fig. 10.31: Efficiency of Solar Cells as a Function of Band Gap [146]

Other concepts to increase the efficiency include using more than one semiconductor in a cell, use of tandem cells, concentrating the sunlight to up to 500 times using lenses, and combining photovoltaic cells with a heat-based technology to use both heat and electricity.

One interesting option is using quantum dots. Their band gap can be adjusted depending on their size and they can produce up to 3 electrons for every high energy photon absorbed. Their band gap can also be adjusted to absorb infrared light and produce electrons. However, several of the materials used in making quantum dots are known to be toxic and the effect of their ingestion in human body is not known. They also degrade easily.

Thin film solar cells have been developed by using chemical vapour deposited or sputtered semiconductors on glass or metals. While these have advantage of using much less amount of material, their efficiency is low. In principle these can be used on windows and be a source of electricity.

Gallium arsenide is a semiconductor material and a compound of Gallium and Arsenic (interconnected cubic lattices). Gallium arsenide (GaAs) has the highest performance of any photovoltaic material, reaching 28.8%. This is possible as GaAs has a direct and more favourable band gap of 1.43 eV — resulting in improved absorption with thinner layers and reduced energy loss. Additionally, GaAs has superior electron-transport properties compared to silicon. Traditionally it is an expensive material and is mostly used in space technologies (satellites/spacecraft).

One of the most successful and widely used thin film solar cells uses a thin film of n-doped cadmium-sulphide and a p-doped cadmium-telluride. Even though cadmium is toxic, with sulphur or tellurium it forms stable crystalline material and remains safe. Cadmium-telluride also has a band gap (1.45 eV) which is close to the optimum value for solar spectrum (1.34 eV) discussed above and thus it works well even in dim light.

CIGS solar cells are also thin film solar cells which are made from a compound semiconductor made of copper, indium, gallium, and selenium. The compound has the chemical formula $CuIn_xGa_{(1-x)}Se_2$ where the value of x can vary from zero (pure copper gallium selenide) to one (pure copper indium selenide). The band gap varies continuously from 1.0 eV to 1.7 eV depending on the value of x, being 1.0 eV for pure copper indium selenide and 1.7 eV for copper gallium selenide. It has a unique property of having a very large absorption coefficient for absorption of light. It also has a great advantage as the semiconductor can be deposited on any flexible base or substrate.

Considerable progress has been made in developing next generation of photovoltaic materials which include dye-sensitised, organic, and perovskite solar cells.

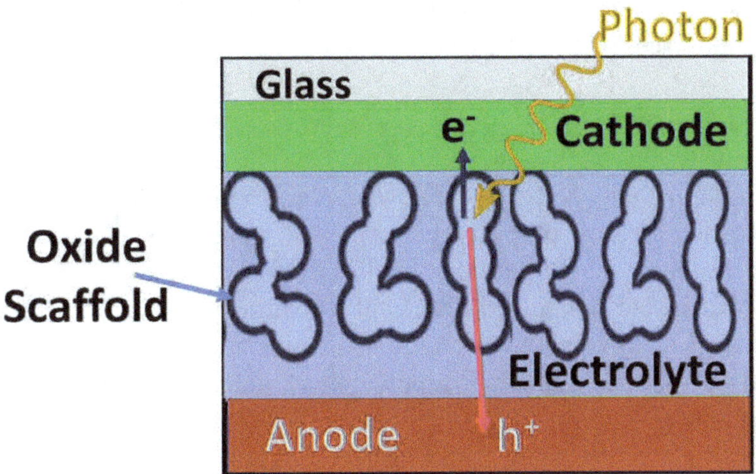

Fig. 10.32: Dye Sensitised Solar Cells [140]

Dye sensitised solar cells use organic dyes to absorb light. The dye is normally coated on titanium oxide which is then immersed in a liquid electrolyte. The dyes absorb the light and electron is released. The electron is transferred to the titanium oxide and the hole is transferred to the electrolyte. Electrodes collect the charge giving rise to current if connected by a conducting wire. These cells are much cheaper than inorganic devices and one can produce them by printing. These can be semi-flexible and semi-transparent. However, these are much less efficient than silicon or other inorganic semiconductors discussed above. The liquid electrolyte can decompose, leak, expand, or freeze. The use of organic compounds can also make them volatile.

Organic solar cells use organic polymers or small organic molecules to absorb light and transport charge to electricity. It is inexpensive to produce these polymers or molecules. Their band gap can be manipulated using molecular engineering and their optical absorption coefficient is high. However, organic photovoltaic cells have low efficiency, low stability, and low strength compared to silicon solar cells.

Perovskite solar cells use perovskite materials like methylammonium lead halides, $CH_3NH_3PbX_3$, where X stands for chlorine, bromine, or iodine — for their light absorbing layer. These are cheap to produce and simple to manufacture. These have excellent light absorption and charge-carrier mobilities resulting in high device efficiencies. However, use of lead raises serious environment concerns. The material is also not very stable.

Work is also in progress for making solar cells using biological materials or specially engineered living cells, which can covert light into electricity.

Advantages and Disadvantages of Solar Photovoltaics

We have discussed only some of the developments in the field of solar cells. As mentioned earlier, because of its immense potential, possibility to have solar power plants of any capacity ranging from a few kilowatts to several tens or hundreds of megawatts and the fact that these do not emit any carbon dioxide during production of power makes them very attractive and a vast literature is available exploring a large variety of materials for their production on an experimental basis or on a large scale.

Solar panels are being deployed on a very large scale for small establishments and remote hamlets. There are suggestions that appliances used at such remote and isolated places may directly use DC current produced by the solar panels and reduce the loss during the AC to DC conversion.

Before closing this discussion, it is important to emphasize several aspects of solar power.

Solar panels have in general an efficiency of about 15–25 percent. However, it is known that their efficiency can get degraded by about one percent every year. Thus, it is expected that they will not remain efficient after some years, requiring their replacement. They occupy large areas and so far, there is no known efficient, safe, and cost-effective method to dispose the non-working panel, which may contain toxic materials. Their fabrication and installation are also very material and energy intensive.

There are fears that lead and carcinogenic cadmium used in the panels can be washed away by rains. In countries like China, Ghana, and India, people living near e-waste dumps often burn the waste to recover copper wires. This requires burning the plastic. The resulting fumes are known to be carcinogenic and cause birth-defects when inhaled. Before long, the world would be producing several tens of millions of tons of waste solar panels every year, for which no cost-effective and safe method of disposal has yet been found. According to some estimates, the world will have 78 million metric tons of solar panel waste by the year 2050. It is hoped that it does not become a menace like the single use plastic today.

The most discussed aspect of solar power is its intermittency and zero power production during night. This necessitates the use of other power generators, like thermal power plants or nuclear power plants to cover the "base load" and the use of "smart grids" employing Artificial Intelligence to handle varying loads. Extensive use of computers in handling this transmission of power exposes them to possibility of cyber-attacks, necessitating use of safeguards.

The panels must also be kept free from dust. A dust level of 4 grams per square meter can reduce the efficiency of solar panels by up to 40%. Keeping the panels clean on a large solar farm is expected to get expensive. Keeping large farms free from dust will also require a considerable amount of water, which may be a problem just where solar farms are most suitably placed, e.g., arid areas or deserts, which also tend to be quite dusty.

One must add that some researchers have suggested using the so-called "lotus effect" to coat the solar panels with a nanostructured hydrophobic surface to keep them free from dust and moisture. This may also raise the cost considerably.

Solar cells work more efficiently at lower ambient temperatures. However, in areas with strong sunlight (in deserts, for example), the temperatures can reach up to 50 degree Celsius, reducing the efficiency of solar cells considerably. Desert areas also tend to have strong wind and dust storms. While the dust storms can cover solar panels with dust, strong winds can damage them.

Even clouds, shades, and bird-droppings can reduce the power produced by solar panels and additional electronic measures (e.g., a bypass) need to be implemented, to address some of these issues.

The occurrences of solar eclipses raise yet another problem. Now that large scale integration of massive solar power farms is becoming routine across the world, the integrated power grids must be prepared to adjust to the decrease and then increase in solar power generation during a solar eclipse. Thus, for example, Karnataka gets 22% of its electricity from solar power. In such events, thermal power plants must be on standby to provide the necessary additional power as the power generation from solar panel drops, to avoid a large-scale power outage.

Considerable work is needed to produce cheap batteries which can store the extra-power produced when the Sun shines brilliantly and for long hours.

One such recent example has been the 100 MW (129 MWh) lithium ion battery installed in James Town Australia, in conjunction with a wind farm, at a cost of about 90 million Australian dollars.

Fig. 10.33: Tehachapi Energy Storage, California [147]

However, we know that batteries are often made from unethically (see later) and unsustainably sourced metals. There have been studies to suggest that if we are to have enough batteries to support renewables meet all the energy needs of the Earth, then the increase in battery production may require four times the metals above what is currently available in the Earth's existing mines and reserves. It will also exceed the amount of cobalt available now and consume up to 80% of lithium on the Earth.

Pumped up hydroelectric plants have been suggested as an alternative plan, where the extra-power is used to pump water to a height and then use it to produce electricity when the production is low. Thus, one can think of pumped up hydroelectric plants as batteries, where (gravitational) potential energy is stored for use at a convenient time. These have the advantage of not using any chemicals. Such systems have a life of up to 75 years and an efficiency of recovery as high as 85% is possible. There have also been suggestions that the water surface on the dams could be covered with solar panels to provide for additional generation of electricity, which can more than compensate for this loss.

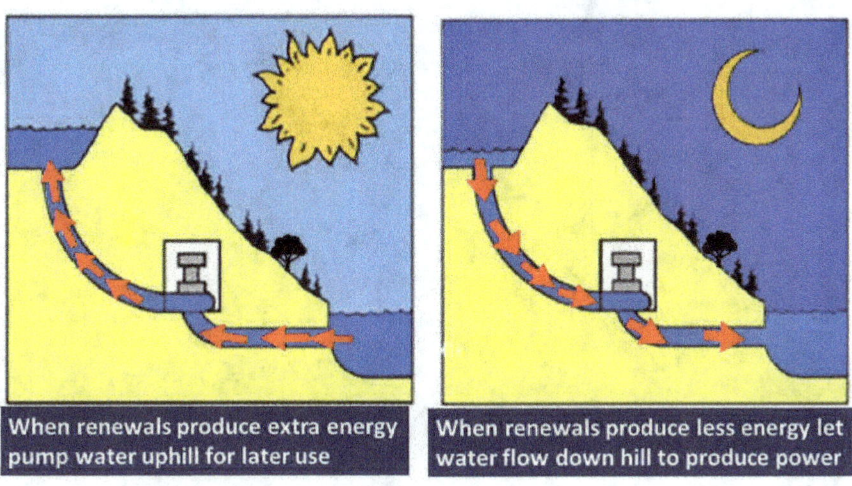

Fig. 10.34: Pumped Up Hydro-electric Storage [148]

There have been suggestions of storing extra electricity produced from solar thermal power plants, or any other power plant for that matter, in the form of hydrogen (produced by hydrolysis of water, see later) to be used at a time of one's choosing. While considering this option, one must remember that hydrogen is highly explosive, and care must be taken for its storage.

Deployment of Solar Photovoltaics

The extensive research in solar photovoltaics and the increasing demand for them across the world has led to a drastic reduction in prices of solar cells and the cost of producing solar power.

This has led to an exponential growth in installation of solar power across the world.

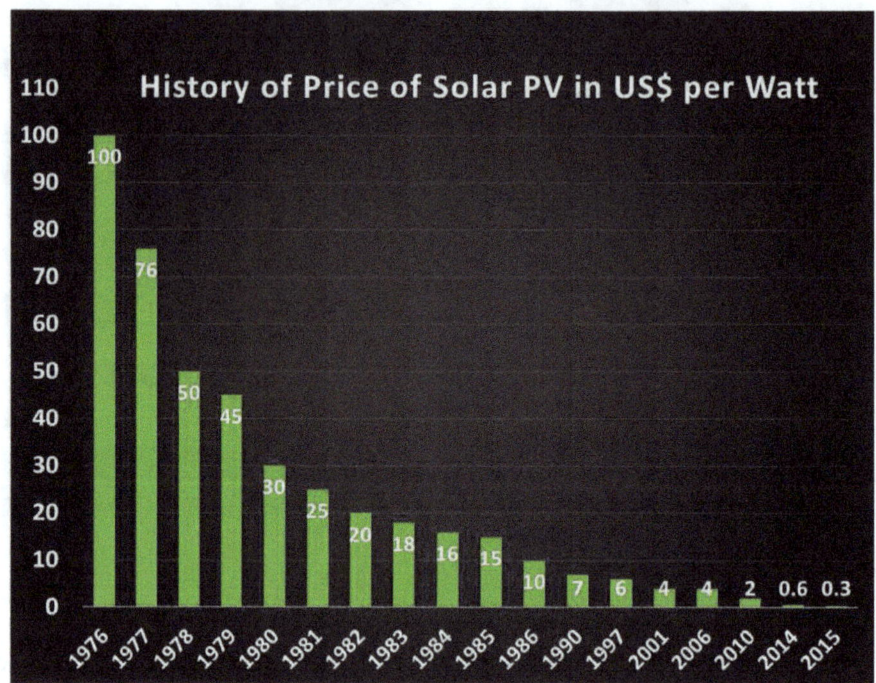

Fig. 10.35: Reducing Cost of Silicon Photovoltaic Cells [149]

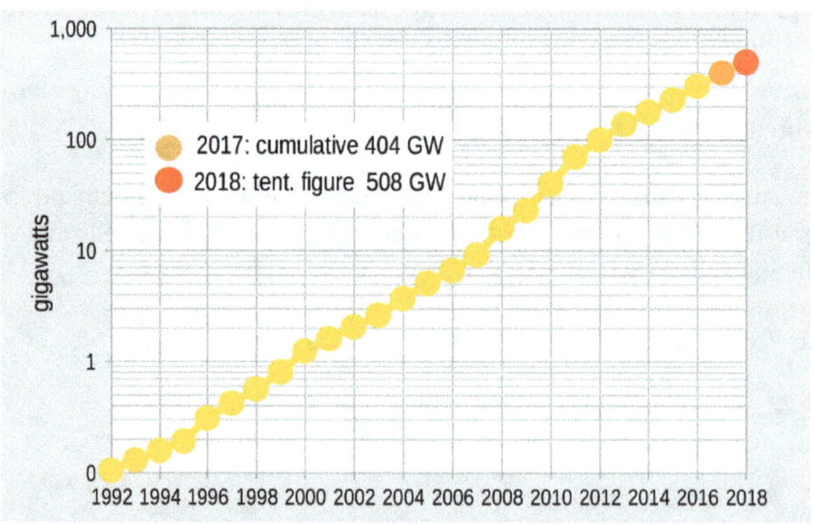

Fig. 10.36: Global Growth of Solar Power [150]

At 26,869 MW in 2018, solar power was 5.4% of total power consumed in India. By June 2020, the installed capacity of solar power in India had risen to 35,120 MW.

Table 10.1: Added and Total Solar Power (MW) in India

Year	Added	Total
2015	2,000	5,050
2016	3,970	9,010
2017	9,100	18,300
2018	10,800	26,869

International Renewable Energy Agency has estimated that the globally installed capacity of solar power could rise by a factor of six over next ten years from a global total of 480 GW in 2018 to 2,840 GW by 2030 and to 8,519 GW by 2050.

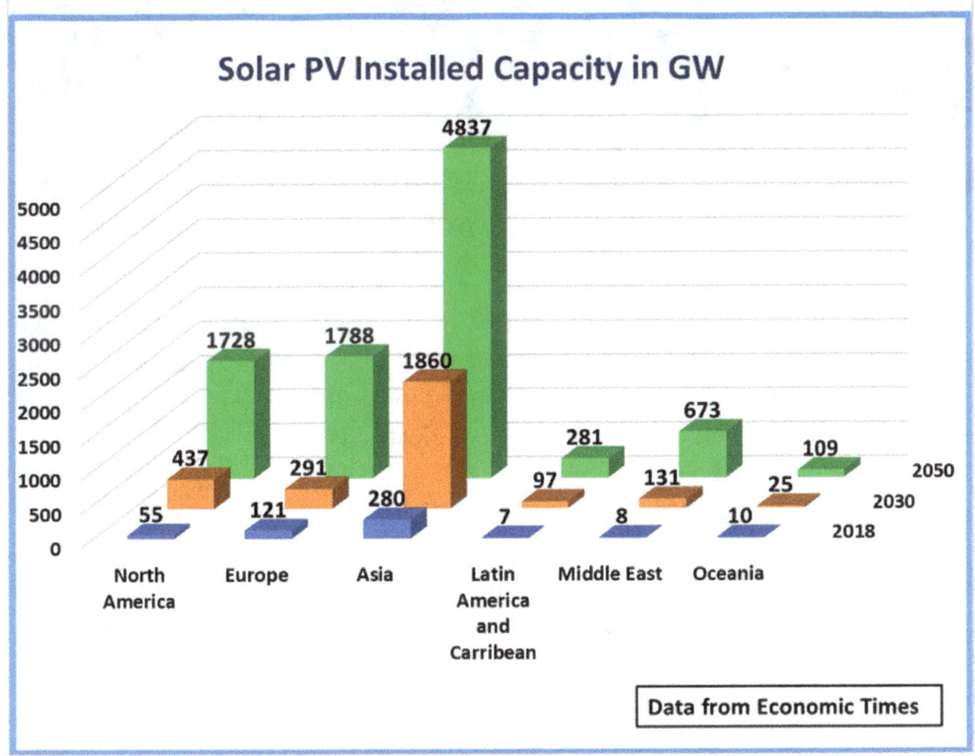

Fig. 10.37: Projections for Installed Capacity of Solar Power [151]

It is also estimated that the levelized cost of electricity for solar photovoltaics will continue to fall from an average of \$0.085 per kilowatt-hour (kWh) in 2018 to between \$0.02 to 0.08/kWh by 2030 and between \$0.014 to 0.05/kWh by 2050.

Fig. 10.38: Pavagada Solar Park (2,050 MW) [152]

We end this discussion by giving a picture of Pavagada Solar Park called Shakti Sthal, in Karnataka, spread over an area of 13,000 acres (53 square kilometres) and built at a cost of Rs. 16,500 Crores. It is slated to be the biggest solar power farm in the world and will produce 2,050 MWs of electricity.

Aquavoltaics and Agrivoltaics

Recall that while discussing hydroelectric power, we mentioned that to avoid evaporation loss of water from canals, as those passed through hot and arid reasons, a cover of solar panels has been used to minimize loss of water by evaporation and to produce power.

It has been suggested that one can grow crops under the solar panels, in an emerging field which has been named "Agrivoltaics".

Fig. 10.39: Cultivation of Betel Leaves in Shade

Several crops, especially leafy vegetables and tubers grow well in shade. Some studies have suggested that leafy vegetables like lettuce and coriander grow larger leaves under the partial shade. Coffee is sometimes planted along with bananas to benefit from the shade provided by the later. One may use solar panels for this.

Fig. 10.40: Solar Panel Integrated Outdoor Mushroom Growing [153]

There have been reports of farmers in Minnesota keeping bees under the solar panels, along with hardy flowering plants.

It is quite common to have lightly shaded enclosures for cultivation of betel leaves in India and Bangladesh. One can use the same area to grow mushrooms and orchids. One could easily replace those shades with solar panels.

Additionally, as one would need to wash the solar panels, for removal of dust — the emerging water can irrigate the plants below. The plants can also keep the solar panels cool by transpiration, as well as reduce the level of dust. In the Sunderbans, women grow tubers like turmeric and ginger on the side of the solar panel, where the water drips.

In several regions of India and the rest of the world, crops are routinely destroyed by hailstorms. Solar panels can provide a protection against this menace. The menace of dust on solar panels should not be underestimated. It is estimated that just four grams of dust per square metre of a solar cell's surface can reduce its energy output by 40 per cent. The water requirements for cleaning of solar panels in India lie between 7000 and 20000 litres per MW per wash. The panels need to be washed every week, for optimum efficiency.

Fig. 10.41: Duel use of Water: Solar Panel Washing and Irrigation [154]

Agrivoltaics proposes to use the fact that most of the agricultural lands across the world generally get adequate sunlight. Thus, putting solar panels above the agricultural forms can help both agriculture and solar power generation. In fact, it is estimated that if we can cover just one percent of the agricultural land with solar panels, we can generate enough power to meet the demands of the entire world.

This will necessitate two changes. Firstly, the height of the solar panels will have to be raised. Normally solar panels are fixed at some angle. With the increase in height they become vulnerable to strong winds. Thus, a mechanism must be incorporated which rotates them to lie horizontally when the wind is strong, to avoid damage.

We have already discussed that floating solar panels can be used on hydroelectric dams to reduce evaporation and loss of water along with several other advantages of the ease of installation.

Amid all these, we would like to remind that rechargeable lithium-ion batteries, which were also used in the 110 MW battery in Jamestown, Australia, mentioned earlier, use cobalt.

Fig. 10.42: Young Children Working in Cobalt Mines in Democratic Republic of Congo [155]

More than 60% of cobalt today is mined in Democratic Republic of Congo, where young children are reportedly made to work in very unsafe and dehumanising conditions. The mining is also destroying pristine forested areas.

Space and Material

We have discussed many aspects of renewable power sources. Before we proceed to discuss nuclear power, we would like to have a closer look at space and materials required to set up various plants. The tables, plots, and numbers quoted for this in the following are taken from Quadrennial Technology Review, Department of Energy, USA published in September 2015 [156]. The plot given later should be seen along with the following tables to get a more complete picture.

Table 10.2: Area Needed for Setting up Power Plants [156]

Energy Technology	**m^2/MW**	**System Boundary Power plant site only; does not consider energy resource mining or collection, processing, or transport area, or used for waste disposal**
Biomass: direct fired	9,000–45,000	Power site only
Coal	270–8,000	Power site only
Coal: Carbon Capture and Storage	12,000	Power site only
Nuclear	6,700–13,800	Low estimate is site only. High estimate includes transmission lines, water supply, and railways for mines, but does not include land used to mine, process or disposal of waste.

Table 10.3: Total Area Needed for Renewable Energy Power Plants [156]

Energy Technology	m^2/MW	System boundary Energy resource extraction area plus power plant site
Biomass: gasification	3,000,000	Site and crop area. Area used primarily driven by biomass productivity and power plant efficiency.
Coal (site and upstream)	40,000	Site and strip mining included.
Geothermal: hydrothermal	1,200–150,000	Low estimate is for site only. Upper estimate includes well-field and plant.
Geothermal: hot dry rock	4,600–17,000	Includes well-field and plant
Hydropower: reservoir	20,000–10,000,000	Site of generators and reservoir
Solar: PV	10,000–60,000	Site of PV system, which includes the area for solar energy collection. PV systems on pre-existing structures have essentially no net increase in land use.
Solar: thermal	12,000–50,000	Site for concentrating thermal systems, which includes the area for solar energy collection.
Wind	2,600–1,000,000	Low-end value is for site only, which includes the physical footprint of the turbines and access roads. The high-end value includes the land area between turbines which is typically available for farming or grazing.

Now let us look at steel, cement, and concrete, *etc.* needed for these plants in tons/TWh, based on the same studies.

We note that the materials needed for solar photo-voltaic (Si) plants include 7,900 tons of steel, 3,700 tons of cement, 2,700 tons of glass, 850 tons of copper, 680 tons of aluminium, 350 tons of concrete, 210 tons of plastic and 57 tons of pure silicon per TWh. For other plants, these requirements are rather small. Thus, windmill farms need 8,000 tons of concrete, 1,800 tons of steel, 190 tons of plastic, 120 tons of iron, 92 tons of glass, 35 tons of aluminium, and 23 tons of copper per TWh. Hydroelectric power plants use 14,000 tons of concrete and 67 tons of steel for the same amount of electricity generation. Nuclear power plants would use just 760 tons of concrete and 160 tons of steel for this.

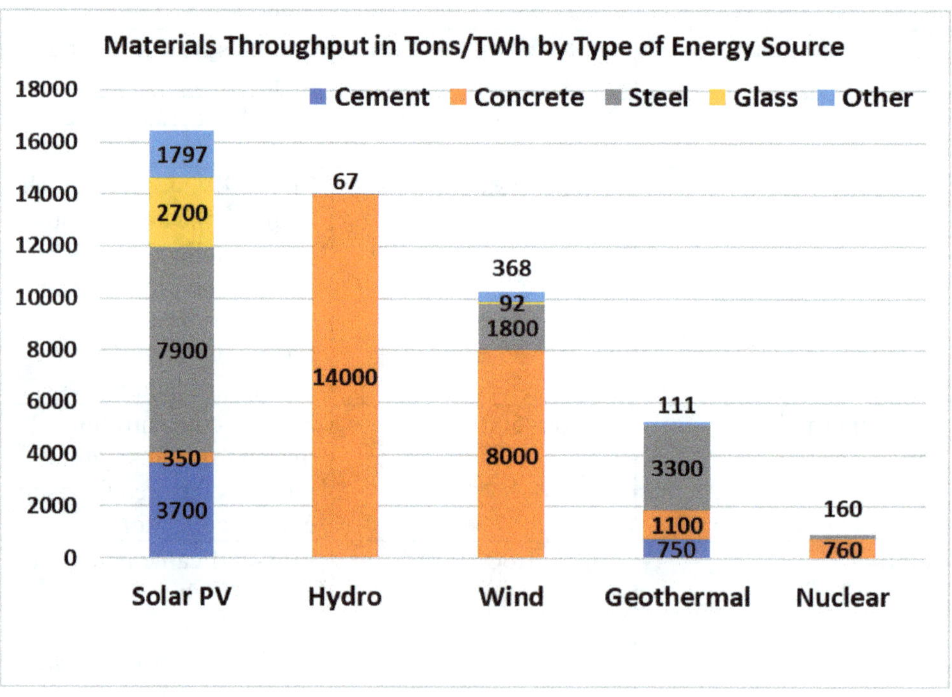

Fig. 10.43: Requirement of Materials for Setting up Power Plants in tons/TWh [157]

The solar panels also need to be changed after 15 to 20 years. The blades of windmills may need to be changed after 20 to 30 years due to tear and wear.

And finally, the solar cells and lithium ion batteries use quite a few rare earths, lithium, and cobalt. Let us look at their continued availability.

These arguments suggest that in near future, many of the materials needed for solar panels may get quite expensive due to difficulty in obtaining them in enough quantity. Some of these, like cobalt, are mined in Democratic Republic of Congo which reportedly employ young children.

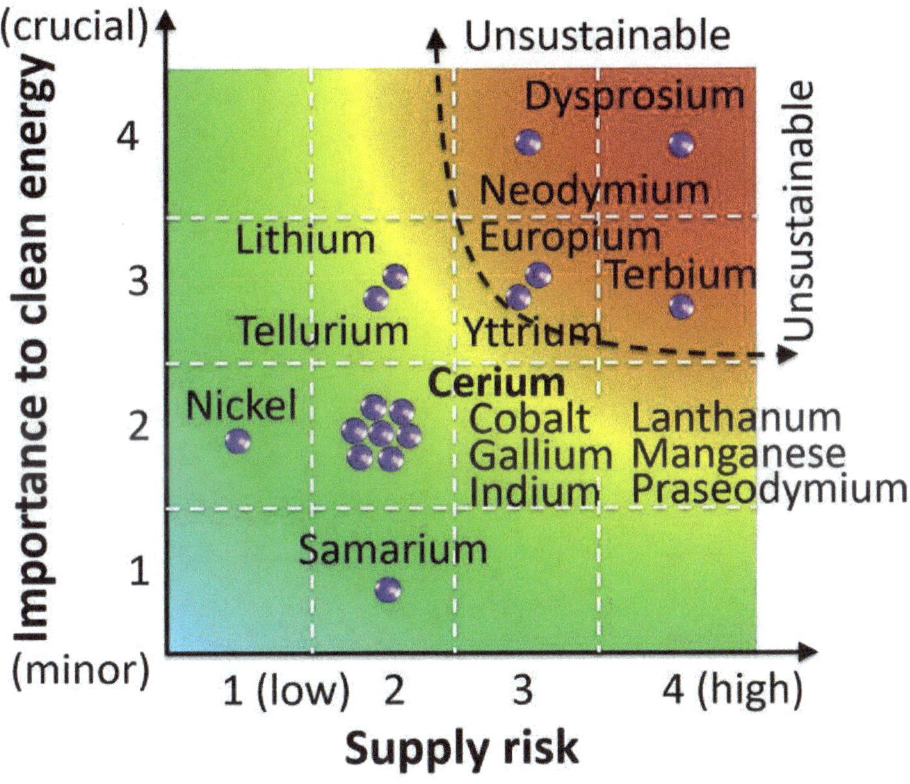

Fig. 10.44: Criticality of Availability of Materials during 2015–2025 [158]

Any long-term plan for energy should take all these factors into account.

Chapter 11

Nuclear Power

Some Basic Nuclear Physics

Before we start to discuss nuclear power in detail, it is worthwhile to recall some basic concepts of nuclear physics. We have already seen that the twentieth century was a turning point in the history of our understanding of the material universe.

We now know that all matter consists of various elements. Atoms, having dimensions of a few times 10^{-8} cm (an angstrom), are the smallest pieces of the elements with definite chemical properties, decided by the unique number and distribution of negatively charged electrons in them. The nucleus consists of an identical number of positively charged protons and (often) varying number of neutrons and resides at the centre of the atom. It has a dimension of the order of a few times 10^{-13} cm (or fermi). Electrons carry (negative) electric charge equal to 1.60×10^{-19} coulombs and a mass equal to 9.1×10^{-28} grammes. Protons have an equal and opposite (positive) electric charge and they are about 1836 times heavier than electrons. A neutron (discovered by James Chadwick in 1932) is marginally heavier (1842 times the mass of the electrons) and neutral. Nuclei having a given number of protons, but different number of neutrons are called isotopes, as the corresponding atoms will have an identical number of electrons, which, as we noted, decides their chemical properties, and they occupy the same position in the periodic table. We also recall that the number of protons in the nucleus is called atomic number, while the sum of the numbers of protons and neutrons is called its mass number.

The energy potential of the nucleus was revealed early in studies of natural radioactivity. In 1896, Henri Becquerel, rather serendipitously, discovered that Uranium salts spontaneously emit penetrating radiations that can be registered in a photographic plate. The name radioactivity was coined by Marie Curie and Pierre Curie. They realized that the new elements Polonium and Radium, which they had discovered, were also radioactive. Thorium and one other new element — Actinium, were also found to be radioactive. The penetrating power of the radiations and their deflections in electric and magnetic fields allowed the scientists to separate three components — alpha, beta, and gamma rays. Alpha decay, beta decay and gamma emission were studied very extensively and used to determine the masses of different nuclei. It is now known that alpha rays are nothing but helium nuclei, beta rays are electrons, and gamma rays are photons.

In the following, we show how U-238 decays to Pb-206 through a series of alpha and beta decays. We now know that decays like these in the interior of the Earth are responsible for its very hot core.

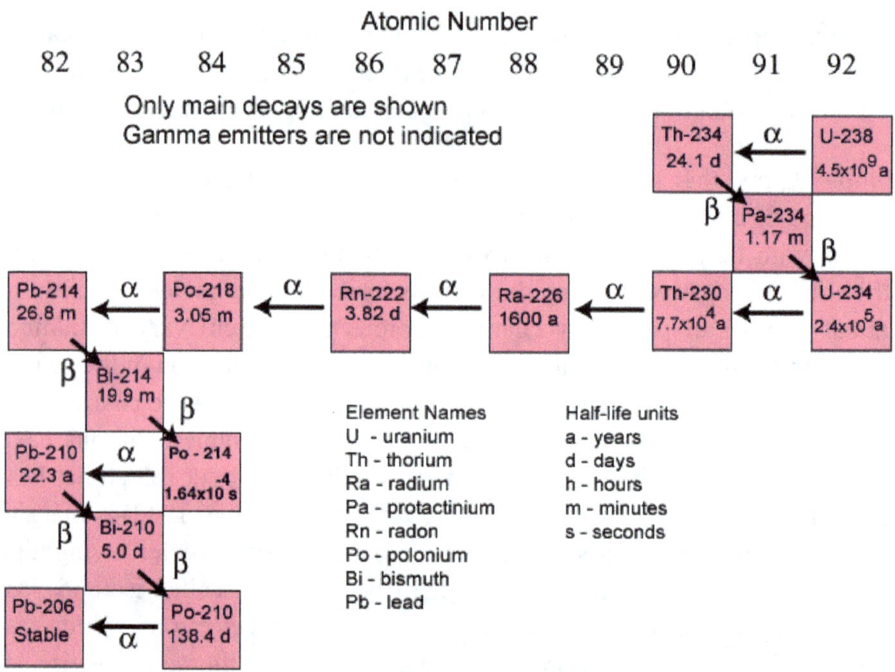

Fig. 11.1: The Decay Chain of U-238 [159]

Next, we look at the chart of all known nuclei (Fig. 11.2). All known nuclei in Nature, fall in a narrow band (along the black squares) in this plot. One side of the band is defined by the neutron drip line beyond which nuclides decay by neutron emission instantaneously. On the other side is the proton dripline beyond which nuclides decay by instantaneous proton emission. Driplines have not been established experimentally for all nuclides even today. The study of formation of all these nuclei in the interior of stars, supernova explosions, during the process of accretion in binary stars, and collision of neutron stars is a fascinating field of research in modern day nuclear physics.

There are no stable nuclides having atomic number greater than 82 (lead), even though bismuth with the atomic number 83 and with a decay half-life of about 2×10^{19} years, is stable for all practical purposes. Thorium-232, Uranium-235 and Uranium-238 are other three naturally occurring elements heavier than lead. However, they are not stable. All of them decay through a series of alpha and beta decays to stable isotopes of lead, each step releasing some energy in the MeV range. It is said that a lump of Uranium is warmer than its surroundings because of the energy release associated with its decay.

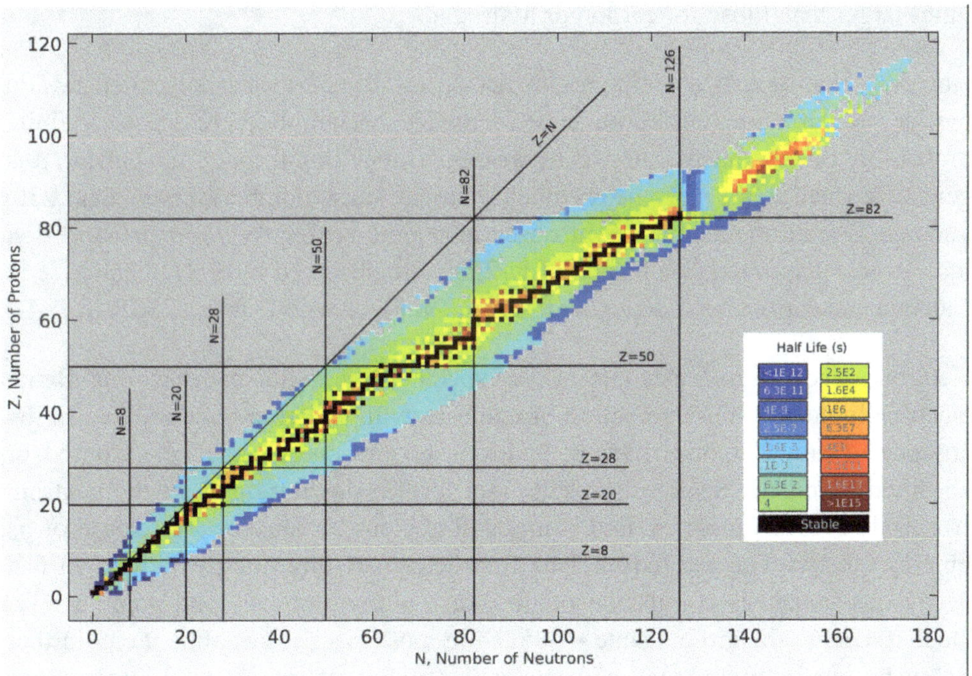

Fig. 11.2: The Nuclear Chart [160]

At this stage let us take a pause and recall that Sir Arthur Eddington, delivering his Presidential Address at the British Association as early as in 1920, argued that the stars must be having sub-atomic reactions as a source of their enormous energy. He further added the prophetic statement: “If, indeed, the sub-atomic energy in the stars is being freely used to maintain their great furnaces, it seems to bring a little nearer to fulfilment our dream of controlling this latent power for the well-being of the human race — or for its suicide". Practical realization of this energy potential of the atom had to wait a few years for the discovery of neutrons, nuclear fission, and the fission chain reaction.

All nuclei having mass numbers more than about 100 energetically favour fission, namely division into two nearly equal parts, the mass of the two fragments being less than the mass of the parent nucleus. In the process involving U-235, for example, a relatively large energy, typically about 200 MeV, is released in the process. It is of interest to recall that in comparison, a typical oxidization reaction like C+2O leading to CO_2 (burning of coal) has an energy release of just 4.1 eV, while 2H+O leading to H_2O (hydrogen burning) has an energy release of about 3 eV per reaction. This provides that the energy density of U-235 is many million times larger than those for carbon or hydrogen.

However, there exists a hindrance to fission — an energy barrier called fission barrier, which varies from nucleus to nucleus. Though there exists a finite probability of spontaneous fission by quantum tunnelling through the barrier, it is generally small. Only very heavy nuclei, having mass number greater than 230, undergo fission spontaneously in observable time scales. We add that the first indications of spontaneous fission of U-238 were observed by S. D. Chatterjee in Calcutta and confirmed independently by G. N. Flyorov in Dubna, USSR in 1941.

Neutron induced fission occupies a very special place in nuclear reactions. Being electrically neutral, the interaction of neutrons with nuclei is quite unique. In the absence of any Coulomb barrier, neutrons of any energy can be scattered or absorbed by nuclei. Scattering results in a further slowing down of the neutron. Absorption results in an excited compound nucleus because of the release of its binding energy. The excited nucleus first decays by any one of the known fast decay mechanisms — emission of nucleons, alpha particles, and gamma rays. Once it reaches the ground state — which in most cases is unstable, it can further decay by one or more steps of beta emission and get transformed into another element with a higher charge number. In the years following the discovery of the neutron, several trans-Uranium nuclei (Actinides) were produced by neutron bombardment and studied in detail. It took some years to realize that fission into two fragments of nearly equal masses is also a competing mode of decay in neutron absorption by heavy nuclei. It was in 1938 that nuclear fission was first discovered by Otto Hahn and Fritz Strassmann and explained by Lize Meitner, in thermal neutron absorption studies by heavy nuclei.

^{235}U was the first nucleus to be identified as thermally "fissile", namely, it can undergo fission under "a gentle tapping" by very low energy neutrons (thermal neutrons), their energy barrier for fission being less than the binding energy of the absorbed neutron.

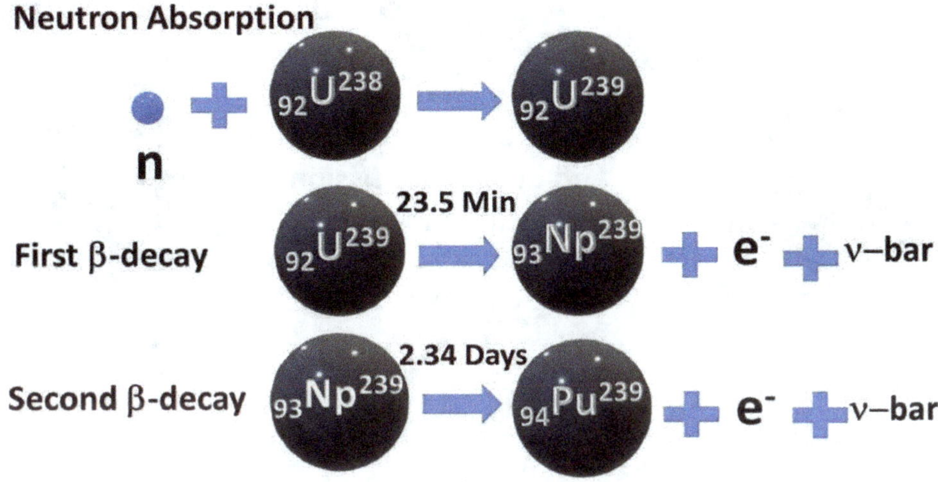

Fig. 11.3: Absorption of Neutron by U-238 and Production of Pu-239

Some nuclei like ^{238}U are not thermally fissile but on absorption of energetic neutrons can either undergo fission or get converted into heavier nuclei. For example, neutron capture by ^{238}U leads to the formation of ^{239}U compound nucleus which can undergo fission or beta decay (with a half-life of about 23.5 minutes) to neptunium-239 (^{239}Np). ^{239}Np in turn undergoes beta decay with a half-life of 2.36 days to produce ^{239}Pu.

Similarly, neutron capture by ^{232}Th produces ^{233}Th compound nucleus which can undergo fission or beta decay (with a half-life of about 22 minutes) to protactinium-233 (^{233}Pa). ^{233}Pa in turn undergoes beta decay with a half-life of 27 days to produce ^{233}U.

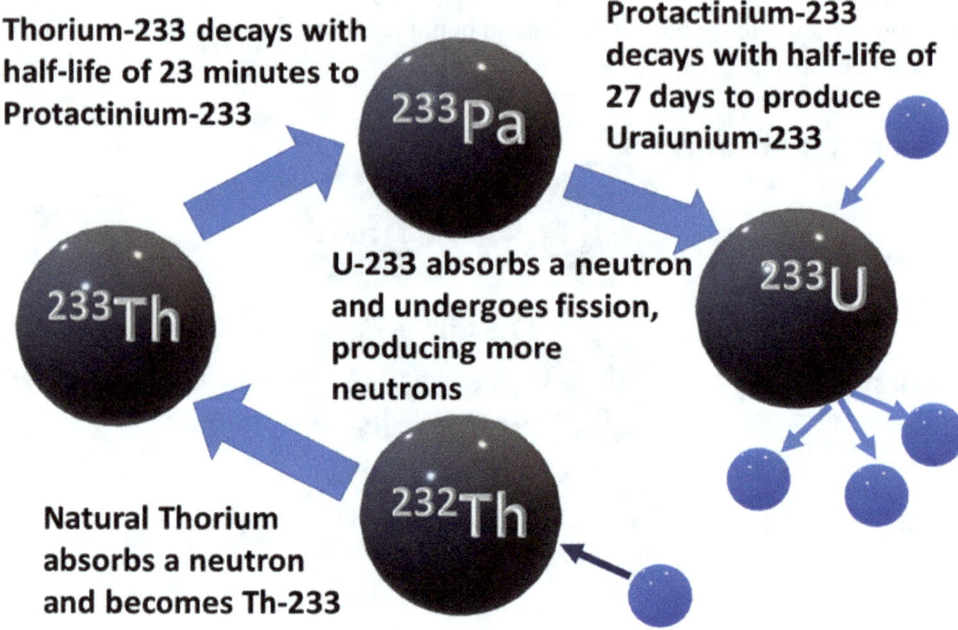

Fig. 11.4: Absorption of Neutrons by Th-232 and Production of U-233 [161]

We also add that neutron absorption cross-section for ^{232}Th is about a factor of three larger than that for ^{238}U. Nuclei like ^{238}U and ^{232}Th are often referred to as 'fertile' nuclei. The process of converting fertile materials into fissile material by neutron capture is often referred to as "fissile material breeding". It is of interest to note that ^{235}U is the only fissile material found in nature in macroscopic quantities. Unfortunately, ^{235}U constitutes only 0.72% of the naturally occurring Uranium ore. ^{239}Pu and ^{233}U are found in nature in trace amounts (along with Uranium and Thorium, respectively produced by absorption of neutrons produced from spontaneous fission). ^{239}Pu is the only man-made isotope which has been produced in hundreds of tons and ^{233}U will possibly be produced in hundreds of tons in future.

A fission reaction, as we have seen earlier, in general results in two fragments of nearly equal masses. The fission fragments not only carry substantial kinetic energy, typically about 200 MeV, but also considerable excitation energy. The fragments lose their kinetic energy to the surrounding medium as heat and lose their excitation energy by de-excitation through the emission of neutrons and gamma rays. In the case of thermal neutron induced fission of ^{235}U, the number of neutrons emitted by the fragments is about 2.5, opening the possibility of a fission chain reaction and sustained energy release. If the produced neutrons can be absorbed by another nucleus which in turn can undergo fission, a chain reaction can be maintained with simultaneous release of a large amount of energy as kinetic energy of the fission fragments. If the chain reaction proceeds uncontrolled, it can result in an explosion.

Fig. 11.5: The Dynamics of Fission on Postage Stamps [162]

One often defines a multiplication factor, *k*, as

$$k = \frac{Neutron\ population\ following\ nth\ generation}{Neutron\ population\ during\ nth\ generation}$$

- If *k* is less than 1, the chain reaction is *subcritical*, and the neutron population will exponentially decay,
- If *k* is equal to 1, the chain reaction is *critical*, and the neutron population will remain constant,
- If *k* is greater than 1, the chain reaction is *supercritical*, and the neutron population will grow exponentially.

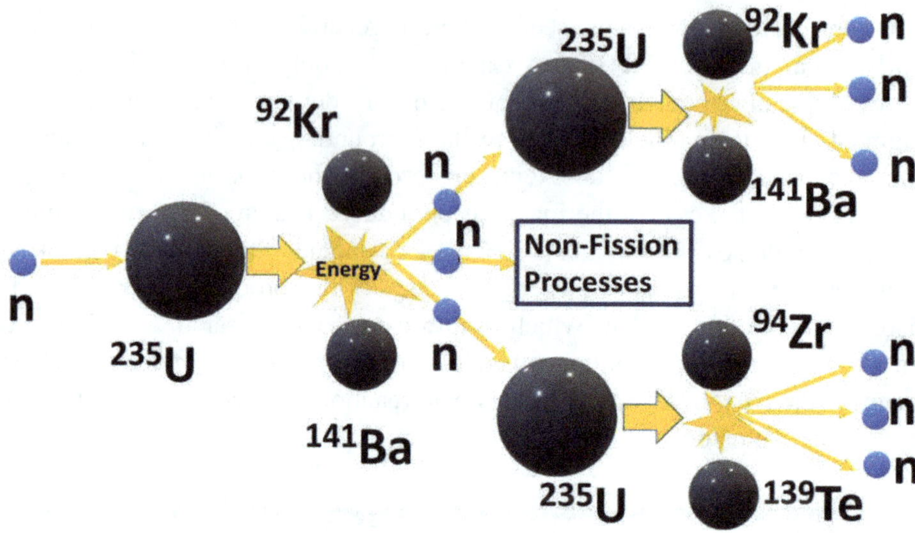

Fig. 11.6: The Fission Chain Reaction [163]

An exponential growth in neutron population also implies an exponential growth in the number of fission events and an exponential growth in energy release. A supercritical assembly of fissile materials in which the neutron population is growing exponentially is a "bomb". On the other hand, an assembly of fissile materials with $k = 1$ can be a perennial source of energy and forms the basis of nuclear reactors.

The Manhattan Project and After

It is an irony that the first demonstration of the enormous energy potential of the tiny nucleus was through its explosive power. The first atomic bomb, made under the leadership of Robert Oppenheimer, was tested on July 16, 1945 ("The Trinity Test"). On this occasion, Oppenheimer is known to have famously quoted from Gita (both in original Sanskrit and English):

"दिवि सूर्यसहस्रस्य भवेद्युगपदुत्थिता।
यदि भाः सदृशी सा स्याद्भासस्तस्य महात्मनः।।11.12।।

If the radiance of a thousand suns were to burst at once into the sky, that would be like the splendour of the mighty one. (Gita: Chap. 11, verse 12)"

Soon after, the United States of America detonated two atomic bombs, 'Little Boy' and 'Fat Man', over the Japanese cities — Hiroshima, and Nagasaki, on August 6 and 9, 1945, respectively. The bombing on Hiroshima resulted in death of up to 237,000 people, most of whom were civilians, directly or indirectly. The bombing on Nagasaki killed up to 80,000 people. The death toll in Nagasaki was lower as the city had earlier been evacuated due to bombings by Allied Forces. It is now known that the 'Little Boy' was a Uranium bomb while the 'Fat Man' was a Plutonium bomb. It was not long after that the then USSR, UK, and France tested their own nuclear weapons in 1949, 1952, and 1960, respectively.

On November 1, 1952, the United States successfully detonated "Mike" the world's first hydrogen bomb, based on the nuclear fusion reaction. It is interesting to recall that, Robert Oppenheimer, Enrico Fermi, and Isaac Rabi had opposed the development of fusion bomb arguing, "Since no limit exists to the destructiveness of this weapon, its existence and knowledge of its construction is a danger to humanity as a whole." Three years later, on November 22, 1955, the Soviet Union tested its first hydrogen bomb. Britain's first successful hydrogen bomb was detonated on November 8, 1957. France tested its first fusion device on August 24, 1968.

China conducted its first nuclear weapon test in 1964 and detonated its first fusion device on June 14, 1967. India conducted its first Peaceful Nuclear Explosion basically a fission device, on May 18, 1974. Its second series of tests including a fusion hybrid device, Pokhran-II, were conducted in May 1998. Within a few days, these were followed by similar tests by Pakistan. South Africa is believed to have tested its nuclear bomb in 1979 and to have dismantled the entire framework afterwards. North Korea tested its nuclear bomb in 2006. Many believe that Israel has nuclear weapon capabilities; however, Israel does not acknowledge it.

Nuclear Reactors for Electricity Generation

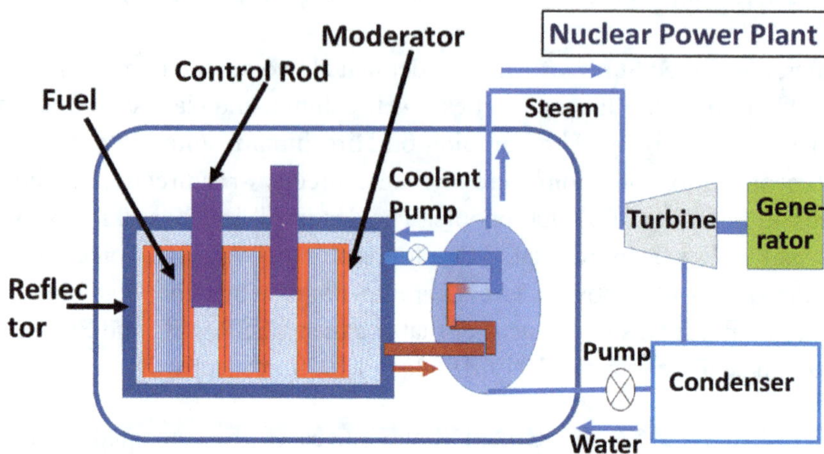

Fig. 11.7: Main Parts of a Nuclear Reactor [164]

A nuclear reactor for power generation is essentially an assembly of fissile and fertile materials arranged in a way that the leakage of neutrons out of the reactor volume is minimized by using suitable reflectors and the probability of their absorption in the fissile core is maximised by slowing down the neutrons using suitable moderators. A coolant transports the heat from the core into a heat exchanger/boiler/turbine system for electricity generation. A set of control rods made of neutron absorbing materials are used to reduce the neutron multiplication factor for routine shut down and in emergencies. Each fission event following the absorption of a neutron by a heavy nucleus in the reactor fuel, on the average, releases more than one neutron by the de-excitation of the two fission fragments. If one of these neutrons is reabsorbed by another heavy nucleus in the reactor fuel causing it to undergo fission, a fission chain reaction will come into operation.

However, the neutrons emitted in the fission event have energies of about 2 MeV. Such energetic neutrons have in general a low cross-section for fission. The fission cross-sections of fissile materials like ^{235}U and ^{239}Pu are very high for very low energy neutrons and decrease rather rapidly with increasing energy of the neutron. Thus, for the chain reaction to continue, the energy of these neutrons must be reduced with the help of "moderators".

Common moderators are normal water (H_2O), heavy water (D_2O) or carbon (in the form of graphite). Neutrons lose their energy by elastic collisions with hydrogen or deuterium or carbon nuclei. Hydrogen in water has the largest moderating power but also a large cross-section for absorption of the neutrons. The moderating powers of deuterium in heavy water and carbon are smaller but their absorption cross-sections for neutrons are also smaller. Many materials such as beryllium, graphite, or steel, *etc.*, primarily reflect the neutrons by elastic scattering rather than absorbing them. These materials can be used as effective neutron reflectors to minimize the neutrons escaping the reactor volume. The energy released in the core can be transferred to a heat exchanger using suitable coolants, as mentioned earlier.

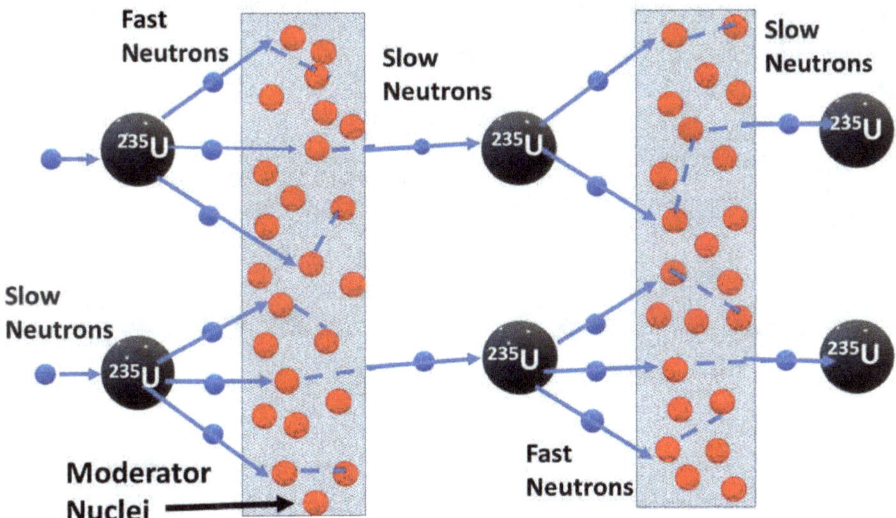

Fig. 11.8: Using Moderator for a Controlled Fission Chain Reaction for U-235 [165]

A set of control rods made of neutron absorbers will have to be used to ensure that the system can be made marginally supercritical to build the necessary power levels in the beginning and maintained at a critical level for steady power generation at the required levels and shut down the reactor when required or in an emergency.

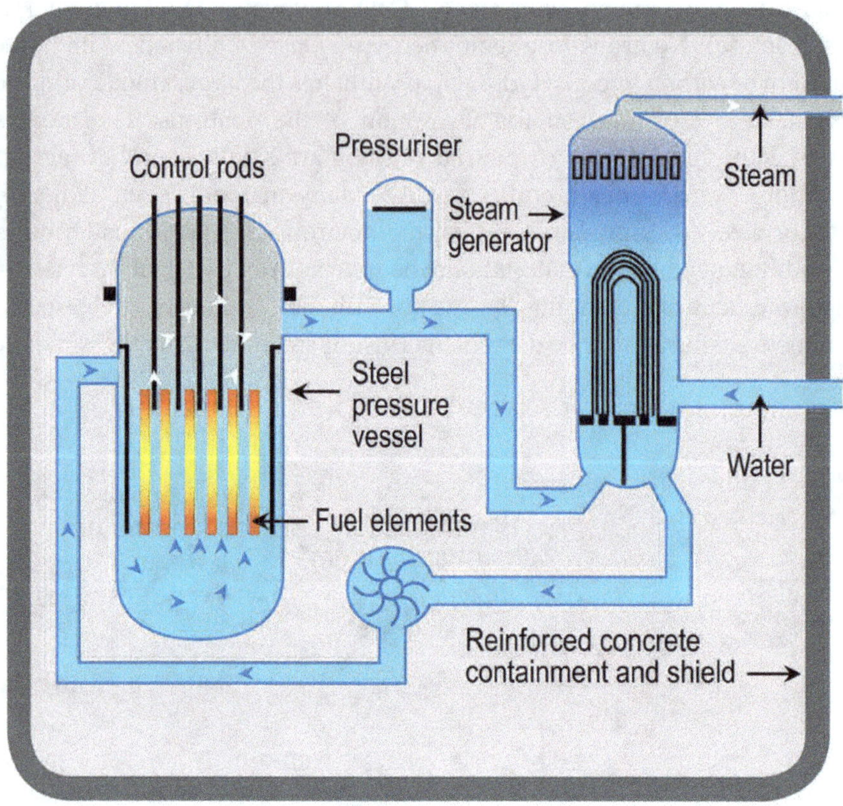

Fig. 11.9: Pressurised Light Water Reactor [166]

Another aspect of neutron emission in fission that plays an important role in the control of the reactors needs to be pointed out here. Neutrons accompanying fission are emitted in two groups — prompt neutrons and delayed neutrons. Prompt neutrons are emitted directly by the fission fragments which are generally formed in an excited state in a time scale of about 10^{-14} seconds. Some of the fragments which are generally neutron rich (delayed neutron precursors) undergo beta decay and a very small fraction of them have enough excitation energy and emit neutrons after the beta decay. These 'delayed neutrons' are emitted at orders of magnitude later times compared to the emission of prompt neutrons and play an extremely important role in the control of the reactors. Power reactors are generally operated in the delayed critical domain and never in the "prompt" critical domain.

The economy of the neutrons in nuclear reactors is shown below.

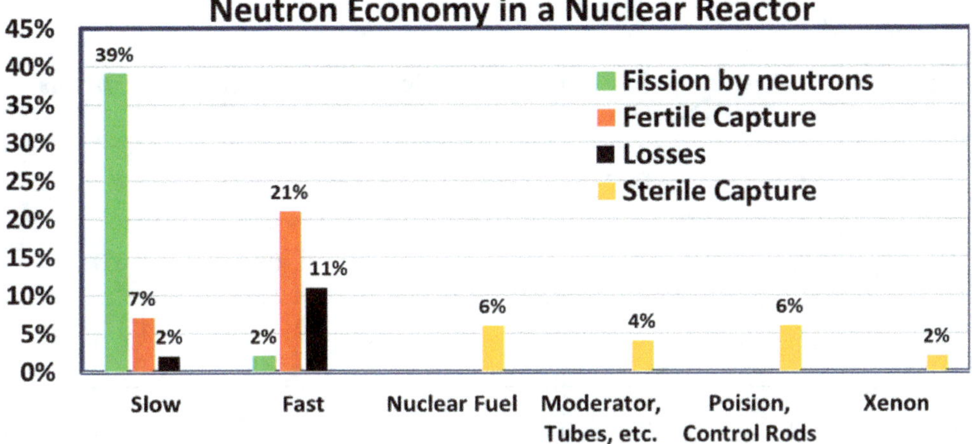

Fig. 11.10: Neutron Economy in a Nuclear Reactor

A typical nuclear power reactor therefore consists of a reactor core housing the fissile material, reflectors to contain the neutrons to the active volume of the reactor and moderators to alter the energy spectrum of the fission neutrons. A circulating liquid or gas enables the removal of heat from the active region and transfer in a heat exchanger and a boiler to produce steam for electricity generation in a conventional turbine. A system of control rods is used to shut down the reactor when required.

Two parameters play important roles in the design of the reactors — the cross-section for fission of the fuel material and the number of fission neutrons emitted following the fission process. We give below the fission cross-sections for thermal (0.025 eV) and fast (2 MeV) neutrons for isotopes of Uranium and Plutonium-239, along with the number of fission neutrons emitted in the process. We would also like to recall that the fission neutrons are in general energetic having energies in the MeV range.

Table 11.1: Fission Cross-Section and Prompt and Delayed Neutrons for Thermal and Fast Neutrons [167]

Isotope	Fission Cross-section (barn)		Prompt Neutrons		Delayed Neutrons	
	0.025 eV	2 MeV	0.025 eV	2 MeV	0.025 eV	2 MeV
U-235	585	1.27	2.42	2.63	0.0162	0.0165
U-238	0.27×10^{-4}	0.57	2.36	2.60	0.0478	0.0478
U-233	531	1.98	2.48	2.63	0.0067	0.0077
Pu-239	747	1.93	2.87	3.16	0.0065	0.0067

We note that Pu-239 has a large fission cross-section for thermal neutrons. Also, the number of neutrons emitted per fission is also close to three. Thus, if the Uranium fuel is left for sufficiently long time in the reactor, Plutonium-239 would be produced and undergo fission as well — inside the fuel, leading to production of energy. The heavy-water moderated reactors can be refuelled without shutdown for years and thus most of the Plutonium is produced and consumed *in situ*! It is estimated that up to 30% of the energy produced in such reactors is contributed by fission of Plutonium. Occasionally a small percentage of Plutonium-239 is added to fresh nuclear fuels.

As was mentioned earlier, natural Uranium ore is primarily ^{238}U with only a small percentage of ^{235}U. ^{238}U does not undergo fission with thermal neutrons. Enriching natural Uranium to have a higher fraction of ^{235}U is technologically complex (see later) and not many countries have access to enrichment technologies. ^{239}Pu is thermally fissile but is not present in nature. It is produced in a reactor core by fast neutron absorption by ^{238}U and must be chemically separated through fuel reprocessing. Again, not many countries have access to fuel reprocessing technologies. Similarly, water, heavy water, and graphite are all good moderators — each one has its own advantages. Consequently, each country has always had preferred strategies and designs of nuclear reactors for electricity generation operating since the mid-twentieth century.

The strategies adopted by different countries were also considerably influenced by their national policies on nuclear non-proliferation. For example, the USA with a large capacity for ^{235}U enrichment has opted for enriched Uranium reactors with light water as moderator. Additionally, USA does not favour fuel reprocessing in the civilian sector to avoid possibilities of ^{239}Pu falling in the hands of "rogue" players. On the other hand, Canada had opted for natural Uranium reactors with heavy water moderators. India, while opting for the Canadian route, has chosen to process the spent fuel to separate ^{239}Pu for its next stage towards migration to its long term three-stage strategy for Thorium utilization in view of its vast Thorium reserves. The safety features like reactor containments, emergency power systems, emergency water cooling systems, *etc.* have been continuously evolving.

The bulk of the world's commercial nuclear reactors belong to one of the many classes of reactors — Boiling Water Reactors, Pressurised Water Reactors, Canada Deuterium Uranium (CANDU) Reactors, Advanced Gas Cooled Reactors, and Vodo-Vodyanoi Energetichesky Reactors (VVER). These reactors are traditionally referred to as Generation II reactors and as mentioned earlier, the choice of the reactor type depended on factors such as fuel availability, national policies on fuel reprocessing, design and manufacturing capabilities and commercial considerations. Gen III reactors are essentially Gen II reactors with evolutionary state of the art improvements in designs and safety features. Gen III reactors have an estimated operational life of about 60 years as compared to 40 years of Gen II systems.

The installed capacities (in GW) of nuclear power plants for several countries are given below.

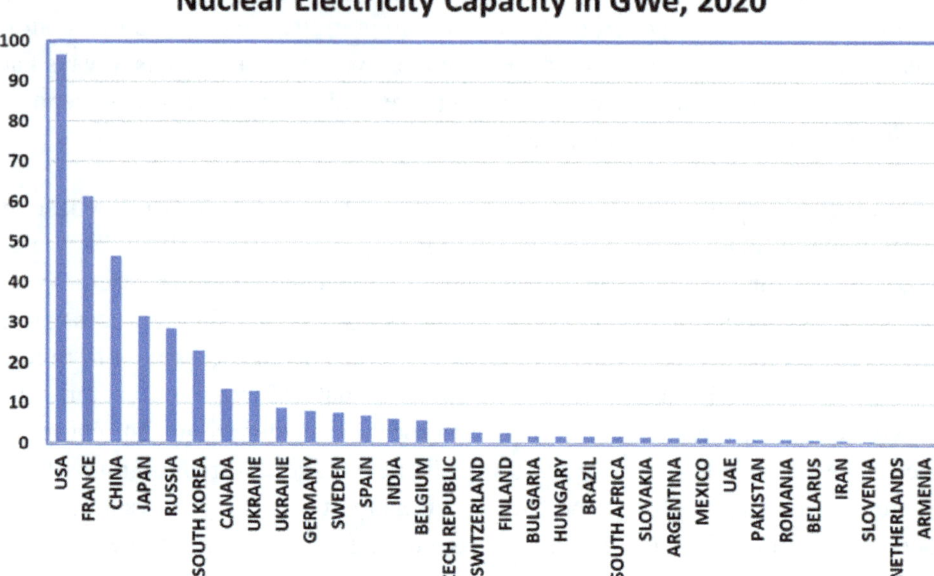

Fig. 11.11: Nuclear Power Capacity of Different Countries in GWe (IAEA, 2020)

Let us look at one other detail, which is often overlooked while discussing Nuclear Power Plants vis a vis other power plants, namely the capacity factor. The net capacity factor of a power plant is ratio of its actual output over a period, to its potential output if it were possible for it to operate at full nameplate capacity indefinitely.

We immediately see that nuclear energy has by far the highest capacity factor of any other energy source. This essentially means that nuclear power plants are producing maximum power for which they are designed more than 90% of the time. Thus, we note that, if we install a nuclear power plant of 1000 MWe, we can be sure of reaching a capacity factor of more than 90%. The corresponding value for a coal fired plant is about 48% and for a solar PV power plant it is only about 25%.

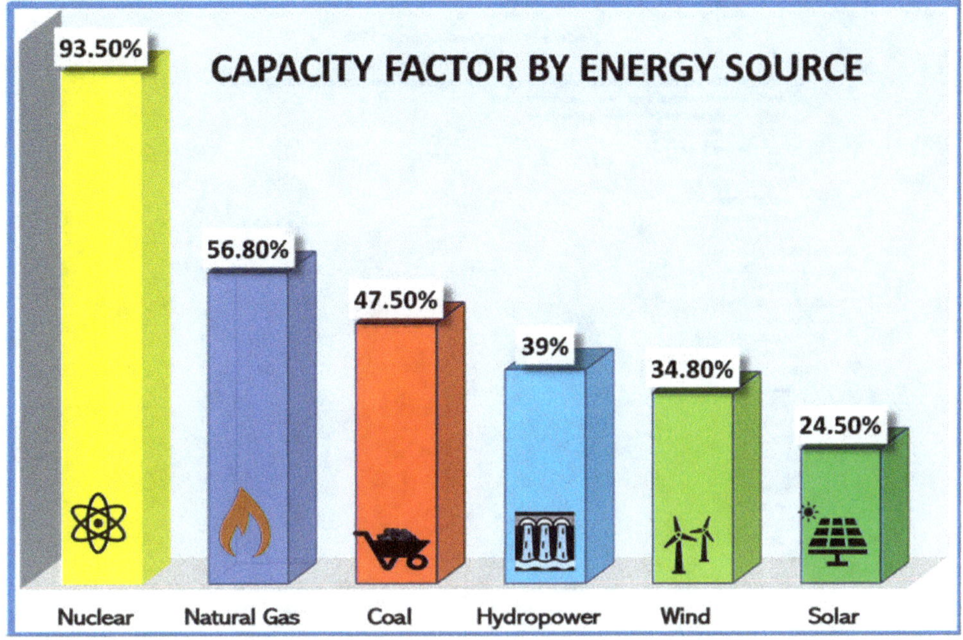

Fig. 11.12: Capacity Factor for Various Power Options [168]

Before closing this discussion, we give the nuclear energy produced by several countries from across the world in a typical year (2019). We also give the contribution of nuclear energy to the overall production of energy across the world in a typical year.

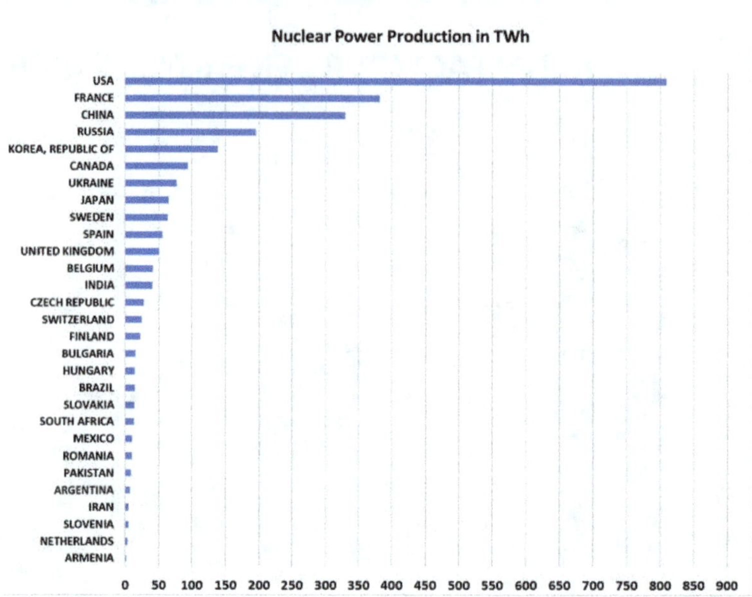

Fig. 11.13: Nuclear Energy Generation by Country (IAEA, 2019) [169]

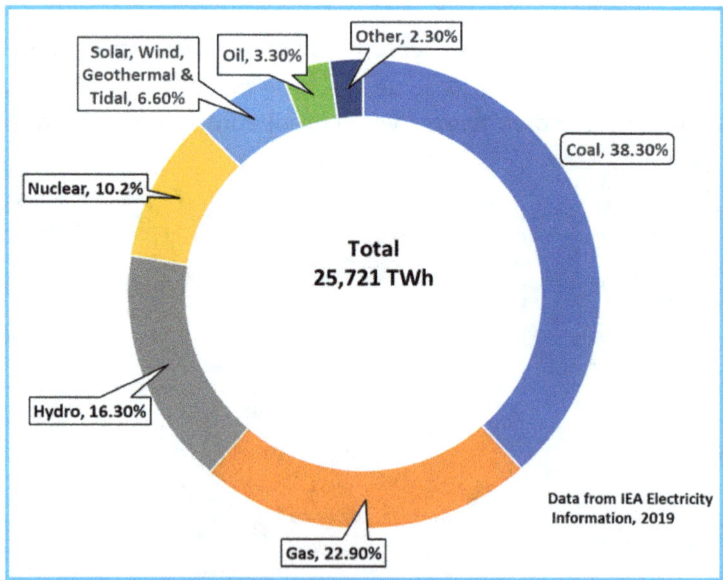

Fig. 11.14: World Energy Production by Source (2017) [170]

It is quite interesting and instructive to look at the growth of nuclear energy globally since its inception.

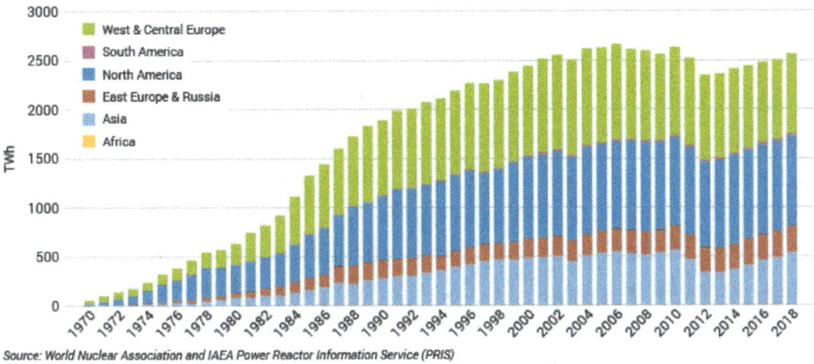

Fig. 11.15: Growth of Nuclear Energy Production with Time]171]

One cannot miss the rapid rise of nuclear energy till almost the 1990's followed by the slowing down thereafter. While there is almost a stagnation since the beginning of the twentieth century with nuclear sources contributing about 10% of the global energy production from all sources, the response of the different countries to the changing perceptions on the role of nuclear energy to the total energy production has not been uniform.

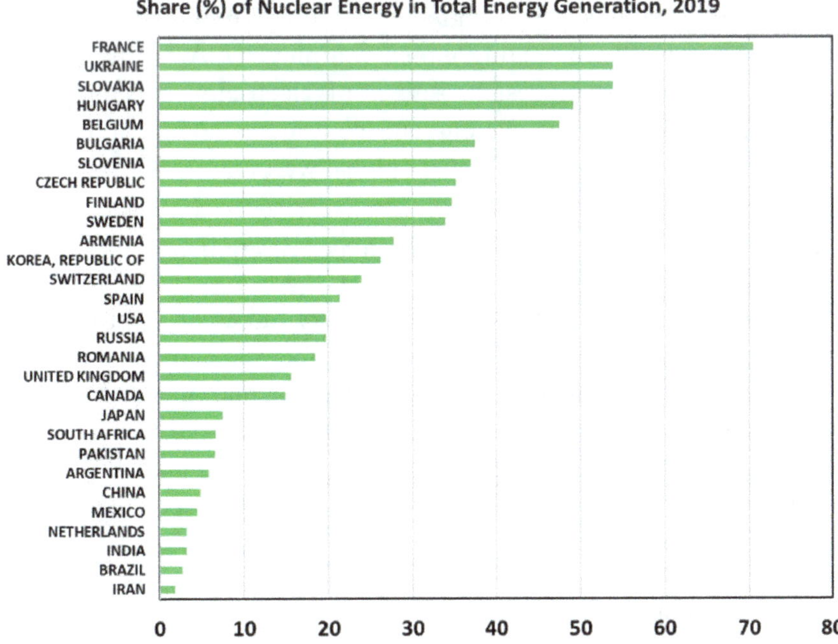

Fig. 11.16: Share of Nuclear Power in Total Energy Generation for Different Countries (IAEA, 2019) [172]

For example, France, Ukraine, Slovakia, and Hungary use nuclear power for catering to 50% or more of the country's electricity demands. China, India, and South Korea are pursuing ambitious expansions of their nuclear power capacities. USA, Canada, United Kingdom, Sweden, and Russia rely on their nuclear capacities to address nearly 20% of their electricity needs. On the other hand, Italy had four operating nuclear power stations for electricity but shut all of them following the Chernobyl accident. Belgium, Germany, Spain, and Switzerland are phasing-out nuclear power. Netherlands, Sweden, and Taiwan have similar intentions. Lithuania and Kazakhstan shut down their lone nuclear stations earlier but plan to build new ones instead. Armenia earlier shut down its lone nuclear plant but then started to utilize it again. Austria built its first nuclear power station but never started to utilize it.

In 2010, before the Fukushima Daiichi nuclear disaster, it was reported that an average of about 10 nuclear reactors would become operational per year. As of 2016, only five came on stream. Global nuclear electricity generation in 2012 was at its lowest level since 1999.

We note that France, with more than 70% of its electricity demand being catered to by nuclear power stations, has essentially zero emission of carbon dioxide. It is a significant step towards mitigating global warming. USA on the other hand, despite being the world's biggest consumer of electricity, gets only 20% of its energy from nuclear power stations and contributes significantly to global emission of carbon dioxide and consequent global warming.

Why Nuclear Power Did Not Grow as Much as Anticipated?

It is an irony that the nuclear industry had to carry the cross of proliferation of nuclear weapons since the very beginning. At the same time, there was an increasing realization that this technology had the potential to satisfy the energy needs of all and cannot remain the privilege of a few. The two International Conferences on the Peaceful Uses of Atomic Energy in 1952 and 1957 in Geneva and the Establishment of the International Atomic Energy Agency in Vienna stand testimony to this collective aspiration of all the countries.

India played a leading role in all these efforts with Homi Jehangir Bhabha presiding over the first Geneva Conference. There were however persistent efforts to limit the technology to a "privileged" few which led to the emergence of the "nuclear club", apparently driven by concerns on the possibility of fissile material falling in the hands of "rogue" nations. There were also concerns on the technical capabilities of processing and storage of long half-life nuclear waste by the new entrants.

Nuclear industry also had a clear division between Eastern Bloc and Western Bloc of countries in 1960s. This led to isolation and lack of exchange of ideas about newer designs with serious consequences including slowing-down of progress. The separate blocs do not exist anymore, but the nuclear industry has not yet become a truly open and competitive international commerce.

The following decades also saw three major accidents in nuclear power stations — Three Mile Island Nuclear Power Station in Dauphin county, Pennsylvania, USA (March 28,1979), Chernobyl Power Plant in Chernobyl, Ukrainian SSR (April 26, 1986), and Fukushima Daiichi Nuclear Power Plant (March 11, 2011) in Okuma, Fukushima Prefecture, Japan over and above several minor accidents. There has also been a recent increase in the fear of global terrorism by non-state actors.

There have been continuous efforts to fortify the Safety and Security of nuclear installations, but these have not come without a cost. The recent discussions on public liability in times of accidents are pushing the cost of nuclear electricity to non-sustainable levels. On the other hand, the costs of renewable energy are decreasing continuously, resulting in a lack of consensus on the role of nuclear electricity in the energy mix to thwart global warming.

The Story of Indian Nuclear Power

India was one of the countries to recognize early the enormous energy potential of the atomic nucleus. As early as in 1944, Homi Bhabha wrote to the Tata Trust that "when nuclear energy has been successfully applied for power production, in say a couple of decades from now, India will not have to look abroad for its experts but will find them ready at home", full one year before the world came to know the power of the atom, full three years before India became free from colonial rule, full ten years before the production of commercial electricity from nuclear energy. The Tata Institute of Fundamental Research, the acknowledged cradle of the Indian Nuclear Programme, came into existence in Bombay in 1945. In March 1946, the Board of Scientific and Industrial Research of the Council of Scientific and Industrial Research set up an Atomic Research Committee under the chairmanship of Bhabha to assess India's atomic energy resources and to suggest ways to harness them. Bhabha's prophetic assessment then, "No power is as expensive as no power" is as valid today as it was then. The Atomic Energy Commission of India was established on August 3, 1948 with Bhabha as its first Chairman. All the research and development activities for nuclear reactors and technology were consolidated in the newly created (January 2, 1954) Atomic Energy Establishment, Trombay (AEET). The Department of Atomic Energy of India was set up on August 3, 1954. The Atomic Energy Establishment was renamed as Bhabha Atomic Research Centre (BARC) after Bhabha's death in January 1966.

Fig. 11.17: Pandit Jawahar Lal Nehru, Former Prime Minister of India Dedicating Apsara Reactor to Nation, January 20, 1957 [173]

One of the early mandates of the Indian nuclear establishment was of course human resource development to implement its ambitious plans for India. The first research reactor to be set up by India was a light water moderated swimming pool reactor, APSARA. Designed by Indian scientists and using slightly enriched Uranium supplied by Britain, the reactor went critical on August 4, 1956.

A second larger research reactor, CIRUS, was built jointly by India and Canada through an intergovernmental agreement under Colombo Plan and went critical in July 1960 using heavy water supplied by United States. Right from its inception, BARC has pioneered almost all research, development and demonstration activities needed for establishing the national nuclear energy programme. An indigenous reprocessing plant was designed and built in 1965. India was one of the few countries outside the recognized weapon states to have mastered this technology at that time. In later years, India also designed and built a research reactor DHRUVA in 1985, which is still used extensively for isotope production and research. Several low power research reactors like ZERLINA, PURNIMA and KAMINI were also designed and built by India for specific purposes.

Bhabha and his team also formulated a three-stage strategy to secure the country's long-term energy independence using the nuclear technologies in the 1950's. The strategy was presented by Bhabha as a part of his Presidential Address during the First Conference on the Peaceful Uses of Atomic Energy in Geneva in 1954. The strategy, basically driven by the known potentials of Uranium and Thorium reserves in the country was also formally adopted by the Government of India in due course.

Let us look at the availability of Uranium and Thorium in the world. India has vast Thorium reserves but only limited deposits of poor-quality Uranium ore.

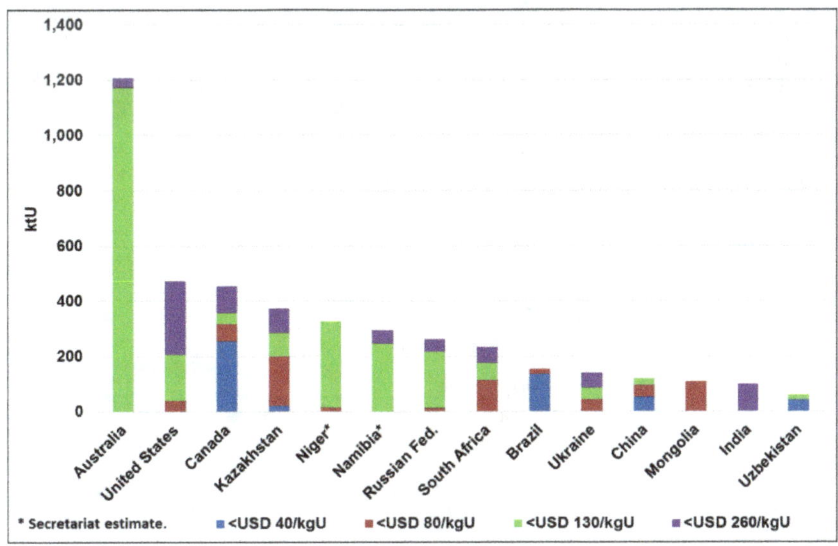

Fig. 11.18: Uranium Resources of the World in kilo tons (Source: IAEA 2018) [174]

India has one of the largest Thorium reserves in the world on its beaches requiring minimum mining. Thorium is also nearly three times more abundant than Uranium in India. It is estimated that 1 ton of Thorium releases about 10 TWh of fission energy. The world thus has enough "easy to mine" Thorium to meet its energy needs for almost thousand years.

Table 11.2: Thorium Reserves in the World]175]

Country	Thorium Reserve (kilo tons)	Country	Thorium Reserve (kilo tons)
India	846	Russia	155
Brazil	632	South Africa	148
Australia	595	China	100
USA	595	Norway	87
Egypt	380	Greenland	86
Turkey	374	Finland	60
Venezuela	300	Rest of the world	1825
Canada	172	World	6355

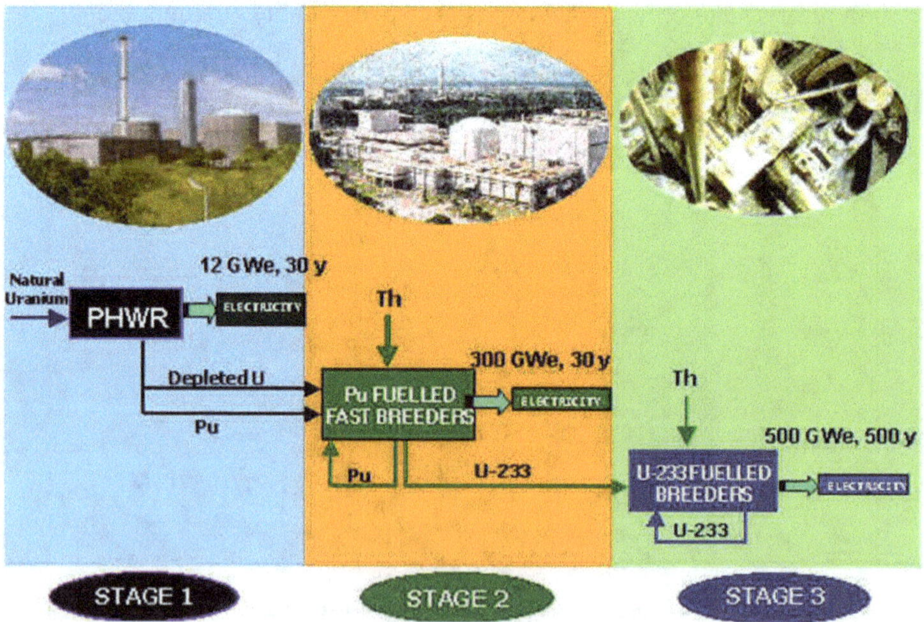

Fig. 11.19: Three Stage Nuclear Power Programme of India [176]

- Stage 1 of the Bhabha's Three Stage Plan was to get the national nuclear energy programme started utilizing the Uranium reserves. As regards the choice of reactor designs, it was India's assessment that pressurized heavy water reactors with natural Uranium fuel to be more suited to the country with no experience in Uranium enrichment at that stage. The by-product, ^{239}Pu of these reactors is key for Stage 2.

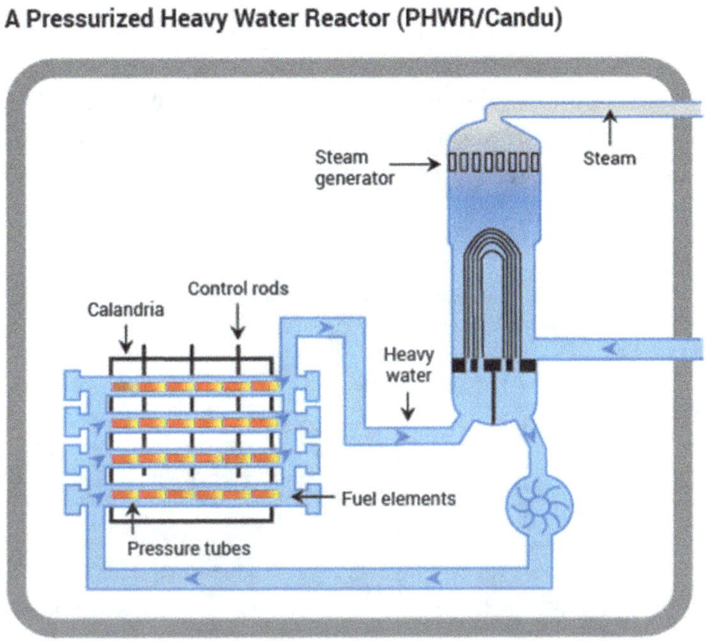

Fig. 11.20: A Typical Pressurised Heavy Water Reactor [177]

- In Stage 2, Fast Breeder Reactors with ^{239}Pu core and ^{238}U and ^{232}Th blankets were to be designed and operated to produce electricity while simultaneously producing more ^{239}Pu and ^{233}U. The large fission cross-section for fast neutrons for ^{239}Pu and emission of more than 3 neutrons per fission makes it an ideal fuel for Fast Breeder Reactors, where more fuel is produced compared to the fissile material used.

The heat removal is normally done using liquid sodium or molten lead. The use of liquid sodium, which is very reactive, poses special technical challenges which have been overcome.

The use of fast neutrons eliminates the need for moderators. This is valuable for breeding of more fuel, as the loss of neutrons by absorption by moderators is avoided and these are available for fissile material breeding.

- In the third and final stage, reactors with ^{233}U were to be designed and operated to produce electricity while breeding more ^{233}U from the Thorium blankets.

We add that Kalpakkam has the unique distinction of being the only place in the world where all the three fissile isotopes *viz.*, U-235 (in Madras Atomic Power Station), Pu-239 (in Fast Breeder Test Reactor), and U-233 (in KAMINI) are used as fuels in reactors.

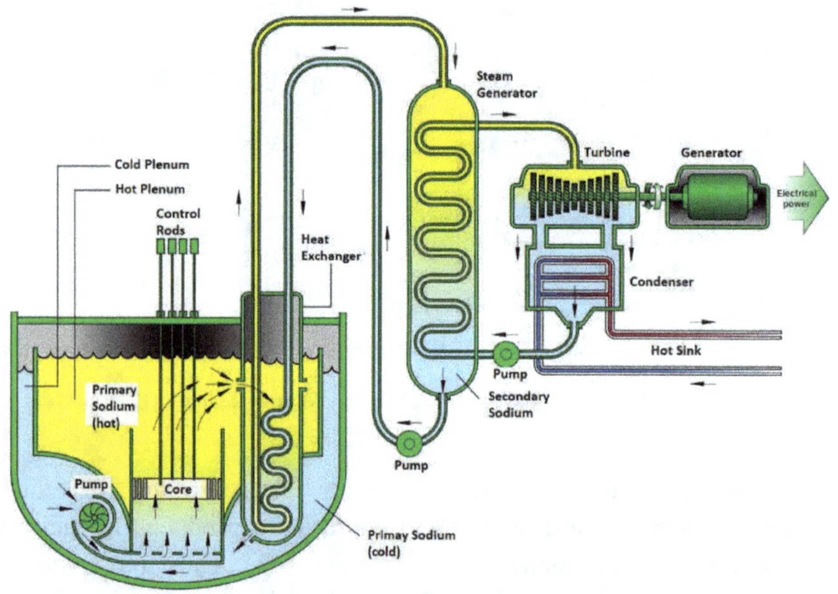

Fig. 11.21: Sodium Cooled Fast Breeder Reactor [178]

Considering the estimates of Thorium deposits in the country, this strategy would ensure enough electrical energy to India for several centuries. The recent discovery of even larger deposits of Uranium in India puts it in even more comfortable situation and gives it additional time to further develop its Stage-2 with Fast Breeder Reactors, *etc.*

In the passing we note that seawater contains about 3 milligrams of Uranium per cubic meter. Considering that total volume of oceans is about 1.37 billion cubic kilometres, this amounts to about 4.5 billion tons of Uranium. It is also believed that this Uranium is continuously replenished as its concentration is controlled by steady state reactions between water and rocks which contain Uranium. Thus, in strictest terms, Uranium is a renewable source of energy. There is a considerable ongoing research on extraction of this Uranium from seawater using fibres coated with amidoxime and made into braids to be anchored to the bottom of the sea. Amidoxime attracts and binds Uranium oxide to the fibres.

Fig. 11.22: Kalpakkam Mini Reactor (KAMINI), India with U-233 as Fuel

Heavy Water and Enrichment of Uranium

Let us also briefly discuss two other important issues — heavy water and enrichment of Uranium, before proceeding.

Ordinary water contains about one deuterium atom for every 6,760 ordinary hydrogen atoms. Until 1943, heavy water was produced by continued electrolysis of hundreds of litres of ordinary water till only a few millilitres of water remained. The residual water is enriched in deuterium content. Now one uses more economical methods like fractional distillation, to get concentrated D_2O which is less volatile than H_2O. Large scale production of D_2O rich water is now mostly done using an isotopic exchange process between hydrogen sulphide and water which produces heavy water over several steps.

India today is one of the largest producers of heavy water in the world, exporting it to countries like South Korea and USA.

The enrichment of Uranium is done using either gaseous diffusion or centrifuge method, both of which exploit the slight difference in the mass of U-235 and U-238. For the gaseous diffusion process Uranium hexafluoride gas is forced through a porous barrier. The molecules of $^{235}UF_6$ penetrate the barrier slightly faster than those of $^{238}UF_6$. The process is repeated till the desired concentration is achieved. In gas centrifuge process a high-speed centrifuge is used to separate $^{235}UF_6$ from $^{238}UF_6$, the former accumulating near the centre.

A potentially useful method uses molecular laser spectroscopy, which utilizes the slight difference in the frequency of light absorbed by molecules having different isotopes of Uranium.

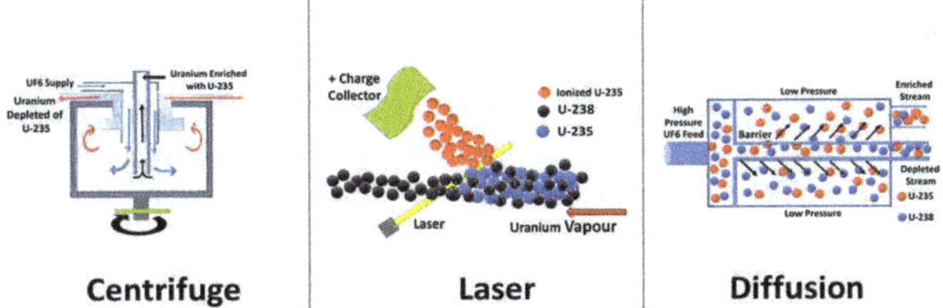

Fig. 11.23: Processes for Enrichment of Uranium [179]

Commercial Nuclear Power in India

The first set of reactors for commercial electricity in India were two units of Boiling Water Reactors of 160 MW each supplied by USA and commissioned in Maharashtra in October 1969. This choice of a light water reactor with enriched Uranium fuel was perhaps prompted by a desire to introduce nuclear power in the country as early as possible and the favourable financial terms offered by US. Discussions with Canada had also been initiated to get two Pressurized Heavy Water Reactors. Design of the reactors and supply of all critical components were the responsibility of Canada. Construction and commissioning were done primarily by Indians. Commercial operation of these two reactors started in 1972 in Rajasthan.

1974 Pokharan Peaceful Nuclear Experiment and the Technology Denial Regime

In May 1974, India conducted an underground nuclear explosion. Though India embarked on the nuclear programme with the sole intention of addressing its energy needs, India had compelling reasons for 'going nuclear', prominent among these being the inequities India perceived in the then emerging nuclear non-proliferation regime with the Non-Proliferation Treaty of 1970 having given a specially elevated status to five nuclear weapon states including China which would have relegated many otherwise capable nations like India to a permanent secondary status. India found this unacceptable and decided to chart out its own course. In a sharp reaction to the nuclear test, both Canada and the US withdrew from their respective collaborations. A US led technology denial regime came into force which denied India not only products but also technology and knowledge basically to limit India's capabilities and capacities in nuclear technologies and in all areas of high technology.

By the way, till 1974 the International Atomic Energy Agency used to conduct regular symposia on applications of peaceful nuclear explosions. This series of meetings was also terminated, with immediate effect.

With determined efforts by the Indian scientific and technological community, unprecedented levels of participation of Indian Industries and the strong political patronage, India not only moved forward in the design, construction, and operation of more PWR reactors but also attained self-sufficiency in heavy water production, Uranium enrichment technology, and complete fuel cycle, *etc.*

It is often joked that the Indian PWRs are no longer CANDU reactors but INDU reactors. A Fast Breeder Test Reactor (FBTR) with mixed oxide fuel was designed and built and has been operating since 1985 in Kalpakkam. Encouraged by the success of indigenising the 220 MWe PWR systems, the Indian engineers also designed and built 540 MWe systems with natural Uranium oxide fuel, moderated and cooled by heavy water.

A 500 MWe Prototype Fast Breeder Reactor (PFBR) with MOX fuel (a mixture of Plutonium oxide and Uranium oxide fuel) is nearing completion in Kalpakkam, building on the decades of experience gained from the lower power Fast Breeder Test Reactor. The PFBR is a pool type reactor with 1750 tons of liquid sodium being used as the coolant.

Fig. 11.24: Fast Breeder Test Reactor, Kalpakkam, India

An advanced heavy water moderated, boiling light water cooled advanced reactor for Thorium utilisation (AHWR) has also been designed as the initial vehicle for our Thorium utilisation programme and will produce as much ^{233}U as it consumes, 75% of the total power coming out of Thorium fuel.

In recent years, India has also gone in for some turn-key reactor systems for power generation from other countries to off-set the slowing down of indigenous additions to the generation capabilities following the post-1974 technology denials. Two VVER systems from Russia are already operational in Kudankulam. Two more are being added. Ultimately, it will have six VVER-1000 nuclear reactors. Negotiations are also on for procuring systems from other countries such as USA and France.

Fig. 11.25: Kakrapar Nuclear Power Plant [180]

Table 11.3: Nuclear Reactors Operating in India

Power Station	State	Type	Units	Total Capacity (MW)
Kaiga	Karnataka	PHWR	220 × 4	880
Kakrapar	Gujrat	PHWR	220 × 2	440
Kudankulam	Tamil Nadu	VVER-1000	1000 × 2	2000
Kalapakkam	Tamil Nadu	PHWR	220 × 2	440
Narora	Uttar Pradesh	PHWR	220 × 2	440
Rawatbhata	Rajasthan	PHWR	100 × 1 200 × 1 220 × 4	1,180
Tarapur	Maharastra	BWR PHWR	160 × 2 540 × 2	1,400

At present India is operating 22 nuclear power reactors with a combined capacity of 6,780 MW. In the following, we give a list of nuclear reactors operating and under construction in India. We also add that nine reactors are under construction, which will bring in additional 6,700 MW of power.

Table 11.4: Reactors under Construction in India

Power Station	State	Type	Units	Total Capacity (MW)	Expected Date of Completion
Kalapakkam	Tamil Nadu	PFBR	500 × 1	500	2020
Kakrapar	Gujarat	PHWR	700 × 2	1,400	2022
Gorakhpur	Haryana	PHWR	700 × 2	1,400	2025
Rawarbhata	Rajasthan	PHWR	700 × 2	1,400	2022
Kudankulam	Tamil Nadu	VVER-1000	1000 × 2	2,000	2026

Additionally, India has an ambitious plan to build 37 more reactors to provide a further capacity of 41,800 MW.

Before discussing newer concepts in reactor design, we add that it is generally felt that small and medium sized reactors can prove to be the most reliable work horses for the immediate future. A standardised and modular design for them can allow mass manufacturing and a drastic reduction in their cost.

The Way Forward: New Concepts in Reactor Designs

We have already noted that for reasons of safety, security and cost, there has been a slowing down of nuclear energy during the recent decades. At the same time, there is a realization that if the world has to contain global warming to acceptable levels as early as possible while at the same time providing enough energy to the global population, there is no alternative but to include nuclear energy in the global energy basket along with other renewable energy sources. Both new reactor designs and manufacturing practices are being examined currently across the world to increase the level of safety and security and to reduce cost. Several new concepts in reactor design have suggested alternatives where it may not be necessary to go through the second stage of Fast Breeder Reactors. One such concept involves molten salt reactor.

Molten Salt Reactor

Molten Salt Reactors use a fluid fuel in the form of very hot fluoride (e.g., LiF-BeF_2-ZrF_4-UF_4; 65%-29%-5%-1%) salt rather than the solid fuel used in most reactors. Since the fuel salt is liquid, it can be both the fuel (producing the heat) and the coolant (transporting the heat to the power plant).

Molten Salt Reactors can accept a variety of fuels like, low-enriched Uranium, Thorium, depleted Uranium, or nuclear waste which is rich in actinides. Lithium fluoride + beryllium fluoride (67%-33%) is used as coolant. It is important that lithium is enriched to minimise lithium-6, which can produce tritium on interaction with neutrons and which could leak, leading to radioactivity.

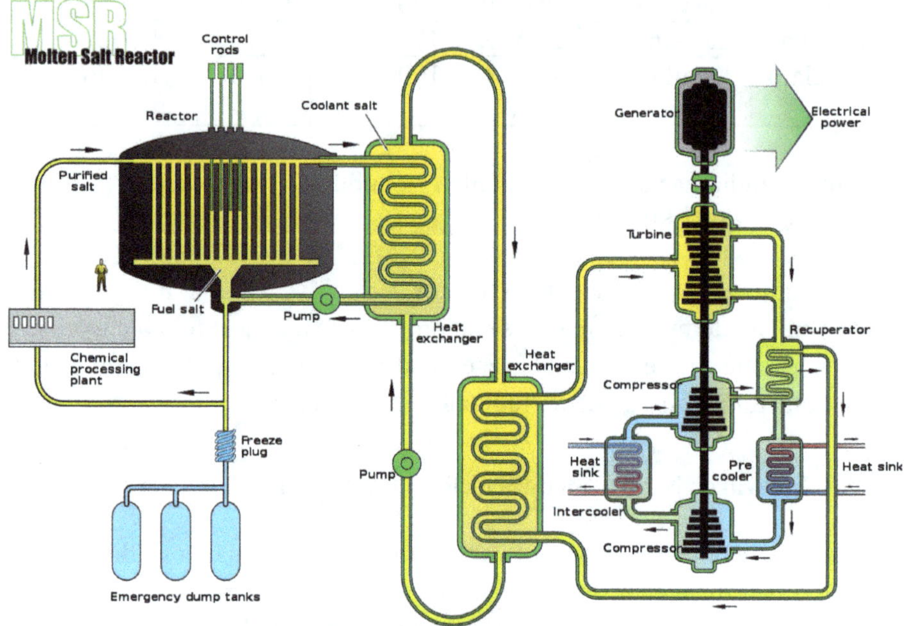

Fig. 11.26: Molten Salt Reactor [181]

The main advantages of Molten Salt Reactors are in the absence of structures like cladding, fuel ducts, grid spacers, *etc.* This ensures that there is no loss of neutrons in these. This helps in increasing the efficiency of fuel and its sustainability. It has also been suggested that one can continuously process the fuel online to remove protactinium, to decay to Uranium-233 and stop it from absorbing additional neutrons. As the boiling point of these molten salts is more than 1400 degree Celsius, one can attain a very high efficiency, for the conversion of heat to electricity.

These reactors allow continuous (online) fueling without shutdown and of course one does not need elaborate scheme for fabrication of fuel pellets, cladding tubes, *etc.*

We do realize that one would be handling highly radioactive, high temperature, corrosive liquids and care must be exercised for the necessary remotely handled chemical plants. Care must also be taken to ensure that the temperature of the salt does not fall below the melting point, else the entire radioactive salt would freeze in the tubes, leading to severe complications. There has been a suggestion to have molten lead as coolant in what is called Dual Fluid Molten Salt Reactor.

Pebble Bed Reactors

One other concept which is being discussed and explored is Pebble Bed Reactor. Some earlier attempts in building and running these were abandoned at some stage due to safety concerns. Recently a prototype for it has been constructed in China and it has been successfully operating for a few years. It is a high temperature gas cooled reactor whose modules can be repeated to have a reactor farm.

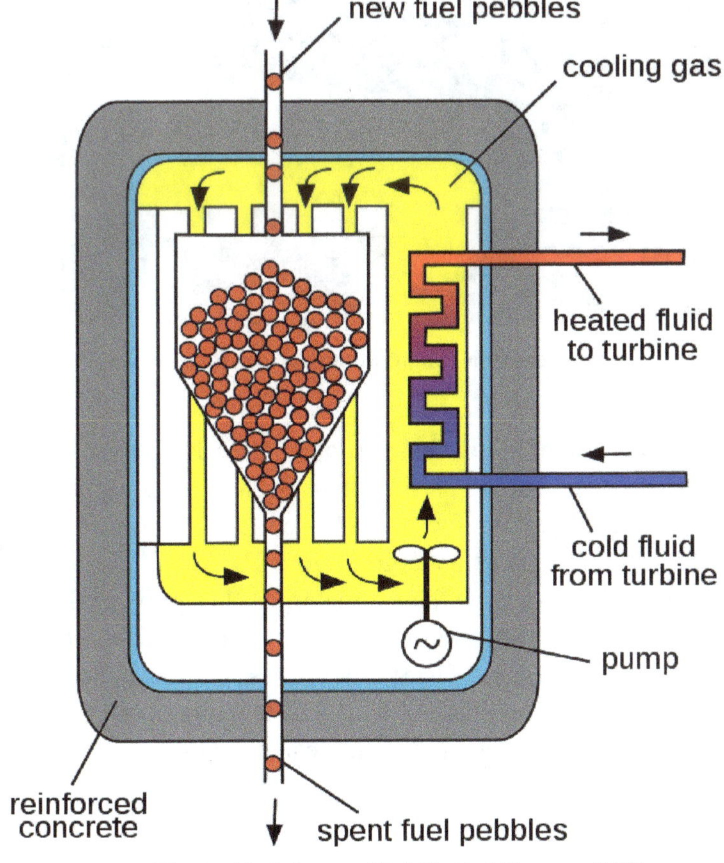

Fig. 11.27: Scheme of Pebble Bed Reactor [182]

The basic design of pebble-bed reactors incorporates spherical fuel elements called pebbles. These are of the size of tennis balls which are made of pyrolytic graphite. The graphite acts as the moderator for neutrons.

Fig. 11.28: Ball for High Temperature Nuclear Reactor [183]

The balls contain thousands of micro-fuel particles called TRISO particles. TRISO stands for TRi-structural ISOtropic particle fuel. Each TRISO particle has a fuel kernel of enriched Uranium. This fuel can also be U-235 with Th-232, so that U-233 produced from Thorium can undergo fission. The kernel is encapsulated by three layers of carbon- and ceramic-based materials that prevent the release of radioactive fission products. The ceramic normally used is silicon carbide which provides a robust integral integrity. TRISO fuels are structurally more resistant to neutron irradiation, corrosion, oxidation, and high temperatures (the factors that most impact fuel performance) than traditional reactor fuels.

As indicated earlier, the triple-coated layers on each of the particles inside the balls act as the containment system and allows them to retain fission products under all reactor conditions. It has been claimed that TRISO particles cannot melt in a reactor and can withstand extreme temperatures, well beyond the threshold of current nuclear fuels. The reactor has thousands of pebbles which create the reactor core, which is mostly cooled by helium gas which does not react chemically with the fuel elements.

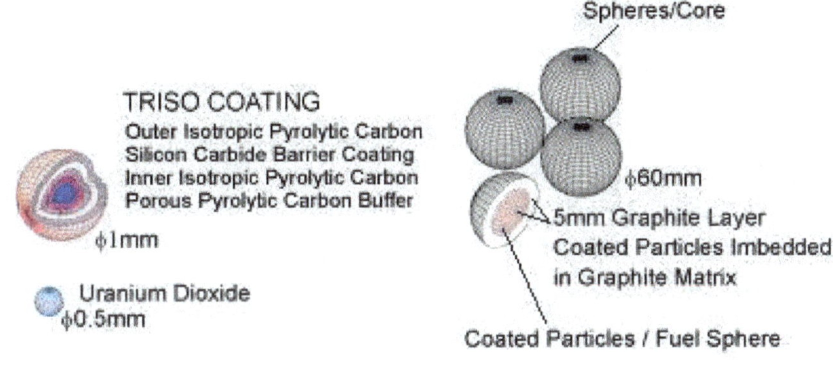

Fig. 11.29: Structure of the Pebble [183]

The reactor is continuously refueled by adding fresh pebbles daily at the top. The older ones are discharged from the bottom of the core. Each pebble can stay in the core for a period of up to three years and pass through the core up to six times for near complete burn up of the fuel.

The spent fuel is then placed directly into dry casks and stored on-site — without the need for interim or active cooling.

Design studies have also been done where pebbles float in a molten salt of lithium beryllium fluoride. The safety issues of pebble bed reactor will be resolved only with enough experience of operating these. These mainly involve radioactivity of graphite pebbles, leakage of radioactive dust, and inflammability of graphite.

It is interesting to recall that Germany had experimented with pebble bed reactors for conversion of Thorium-232 to Uranium-233. However, the integrity of the TRISO fuel cells was so good that it became very difficult to extract Uranium-233 from there!

Generation IV International Forum (GIF)

Gen IV reactors are a set of nuclear reactor designs currently being researched for commercial applications by the Generation IV International Forum with technology readiness levels varying between a level requiring a demonstration to economically competitive implementation. The Forum, founded in 2001, has fourteen members out of which nine are founding members.

Argentina (non-active)

- Brazil (non-active)
- Canada
- France
- Japan
- The Republic of Korea
- The Republic of South Africa
- United Kingdom
- USA
- Switzerland
- Euratom
- People's Republic of China
- Russian Federation
- Australia
- European Union

The Forum has identified several nuclear energy systems for further development, based on a variety of reactors and fuel cycles technologies. Their designs include thermal and fast neutron spectra cores, closed and open fuel cycles. The reactors range in size from very small to very large. Molten Salt Reactor discussed above is one of the designs to be studied. Improved safety, sustainability, reliability, efficiency, economics, proliferation resistance and physical protection are some of the driving forces of the research efforts of the Forum [184].

It is unfortunate that GIF is not a truly international effort, for example under IAEA. India is not a member of this initiative for historical reasons even though India is recognized as a major player in the nuclear field in the coming decades.

There is no International platform to implement safety and security of nuclear installations as in civil aircraft operations, which has been operating for decades with excellent results.

Chapter 12

Nuclear Fusion

The Sun and all the stars in the universe, derive their energy from fusion reactions. The sunlight and the heat from the Sun, along with water and oxygen on Earth have led to evolution of life on our unique planet, as we saw earlier.

Fig. 12.1: A Part of the Digitised Sky with the Nearest Star Proxima Centauri (in red) [185]

The most common fusion mechanism in stars proceeds as follows: First, two protons fuse to make a deuteron, giving out a positron and a neutrino. In this process, 1.44 MeV of energy is released and it accounts for 10% of the toal energy produced in the Sun. Next, the deuteron absorbs a proton to make helium-3 and radiate a gamma ray. In this process 5.49 MeV of energy is produced and it accounts for close to 40% of the energy produced in the Sun. Next, two helium-3 nuclei fuse, release two protons and produce helium-4 along with 12.86 MeV of energy. This process accounts for 39% of the energy produced in the Sun. This sequence is call "p-p Chain I" and takes place almost 90% of the time in proton-proton collisions. The rest of the energy is produced by other (fusion) reactions.

However, for these reactions to proceed one needs temperatures of the order of millions of degrees Celsius as protons and other nuclei are positively charged and their Coulomb repulsion opposes their coming together. They have to have enough kinetic energy to overcome the Coulomb replusion and get within the reach of the nuclear attraction which extends only up to their radii. The high temperture provides the neccesary kinetic energy to the fusing nuclei.

The core of the Sun has a temeprature of about 15 million degree Celsius. The cross-section for the proton-proton fusion reaction in the chain discussed above is extremely small (as it proceeds via a weak process) even at such high temperatures (energies). The occurrence of the reaction is helped by the enormous density and volume of the protons in the interior of the Sun, forced by the intense gravity of the Sun.

For laboratory production of energy using fusion reactions, several other fusion reactions having much larger cross-sections are considered.

1. ${}^2_1H + {}^3_1H \rightarrow {}^4_2He$ (3.5 MeV) + n (14.1 MeV),

2. ${}^2_1H + {}^2_1H \rightarrow {}^3_1H$ (1.01 MeV) + ${}^1_1H$ (3.02 MeV),

3. ${}^2_1H + {}^2_1H \rightarrow {}^3_2H$ (0.82 MeV) + n (2.45 MeV),

4. ${}^2_1H + {}^3_2He \rightarrow {}^4_2He$ (3.6 MeV) + ${}^1_1H$ (14.7 MeV).

The first reaction in the above corresponds to the fusion of deuterium (D) and tritium (T) and has the largest fusion cross-section of about 5 barns (recall that 1 barn = 10^{-28} sq metres) of all the potential fuels considered above, and it reaches the lowest centre of mass energy of about 65 keV.

For this reason, D-T fusion is the most common reaction which is being attempted in the laboratory for several decades now.

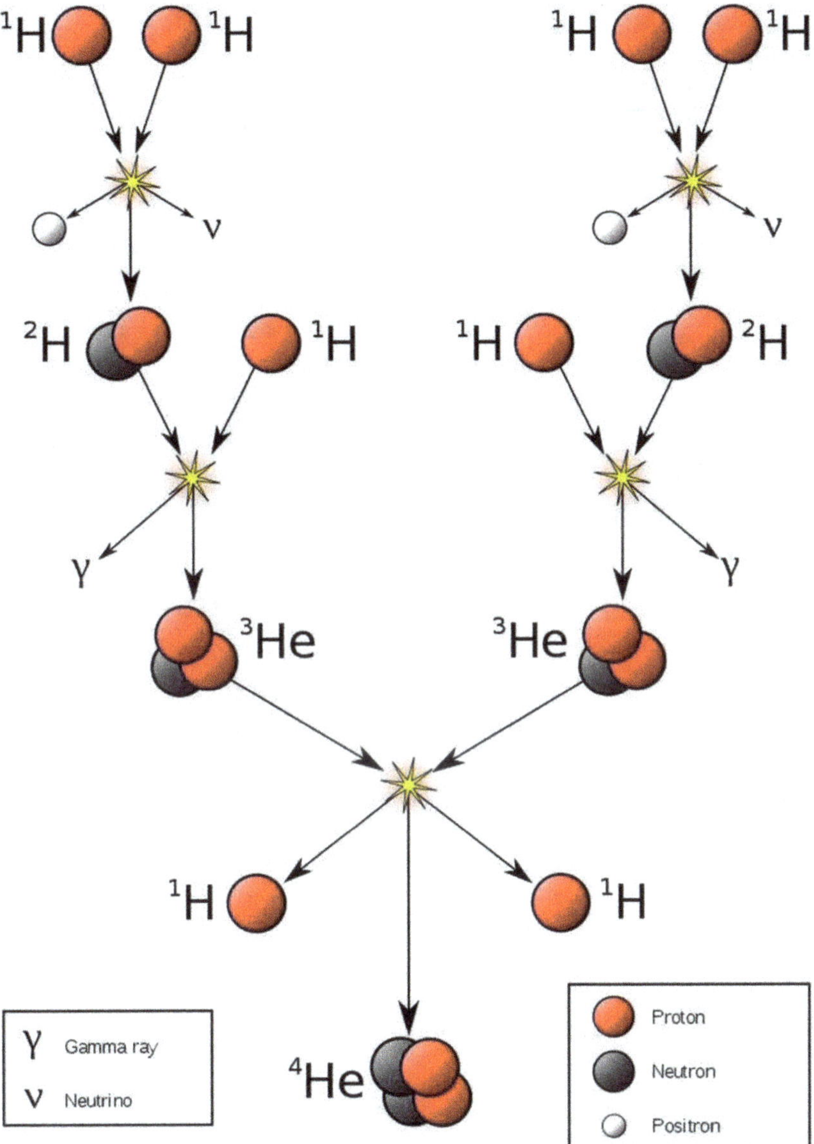

Fig. 12.2: Fusion of Protons Leading to Production of Energy in Sun and other Stars [186]

Deuterium is quite common, being one atom for 6420 atoms of hydrogen, and one can see that we have enourmous supply of it in our oceans and we have already separated several thousand tones of heavy water (D_2O), where hydrogen is replaced by deuterium.

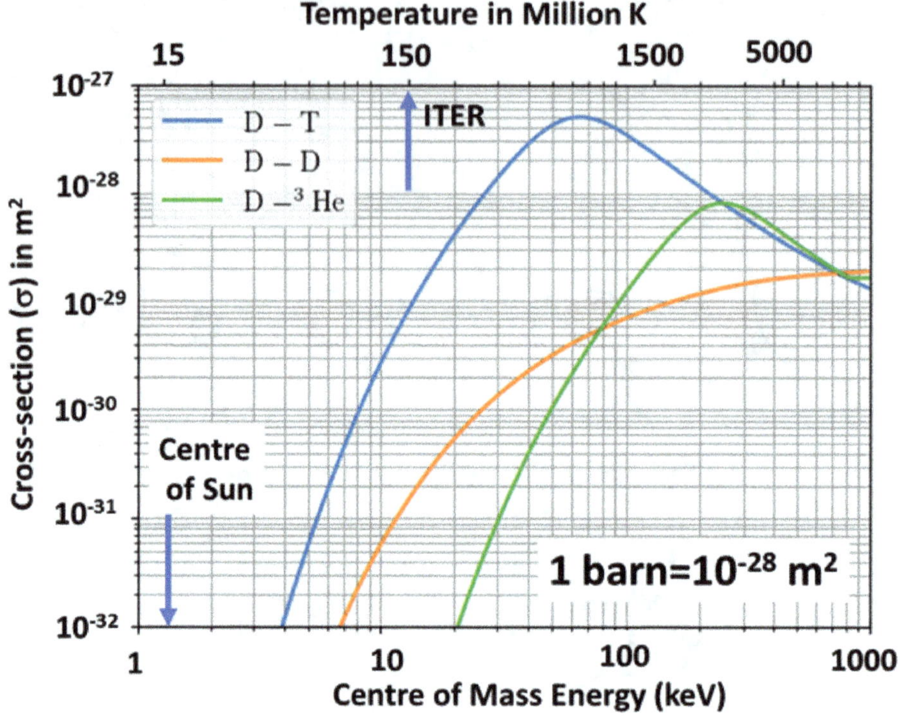

Fig. 12.3: Cross-section of Some Fusion Reactions, Including Those for DT Fusion [187]

Tritium is radioactive, with a half-life of 12.5 years, and its only trace amount is found in the atmosphere, though heavy water used in PHWR will develop some concentration of tritium.

It is mostly produced by action of slow neutrons on lithium-6:

${}^{6}_{3}Li + n \rightarrow {}^{4}_{2}He$ (2.05 MeV) + ${}^{3}_{1}H$ (2.75 MeV).

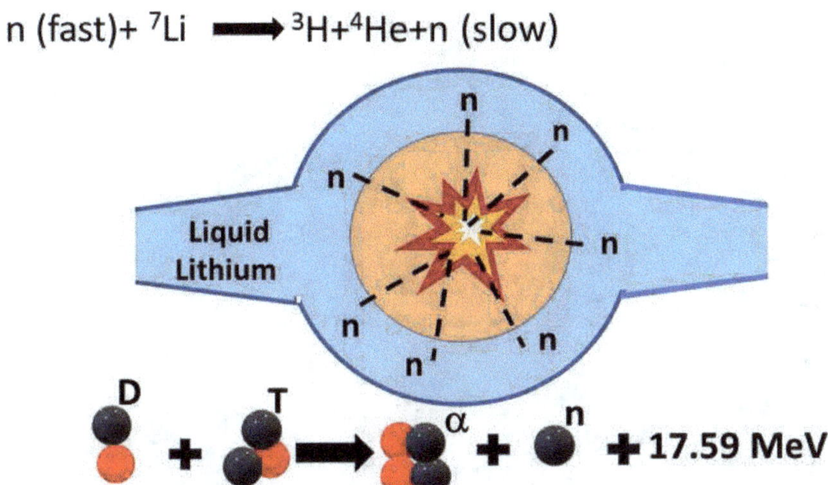

Fig. 12.4: Breeding of Tritium Using Fast Neutrons [188]

Thus, the promising schemes are attempting to have a DT fusion, followed by absorbtion of neutrons in a blanket of lithium-6 to produce tritium. The blanket can also have Thorium-232 to convert it to Uranium-233.

Tritium breeding can also be done by action of fast neutrons on lithium-7:

${}^{7}_{3}Li$ + n (fast) → ${}^{4}_{2}He$ (2.05 MeV) + ${}^{3}_{1}H$ (2.75 MeV) + n(slow).

Simplicity of the scheme, easy availibility of the fuel, absence of any radioactive waste, and a large gain in energy in these reactions give us a vision of plentiful supply of energy for billions of years to come. These expectations gave rise to enormous excitement till it was realised that to contain the hot plasma at a very high density for a sufficiently long time to ignite a fusion reaction and to sustain it is quite difficult.

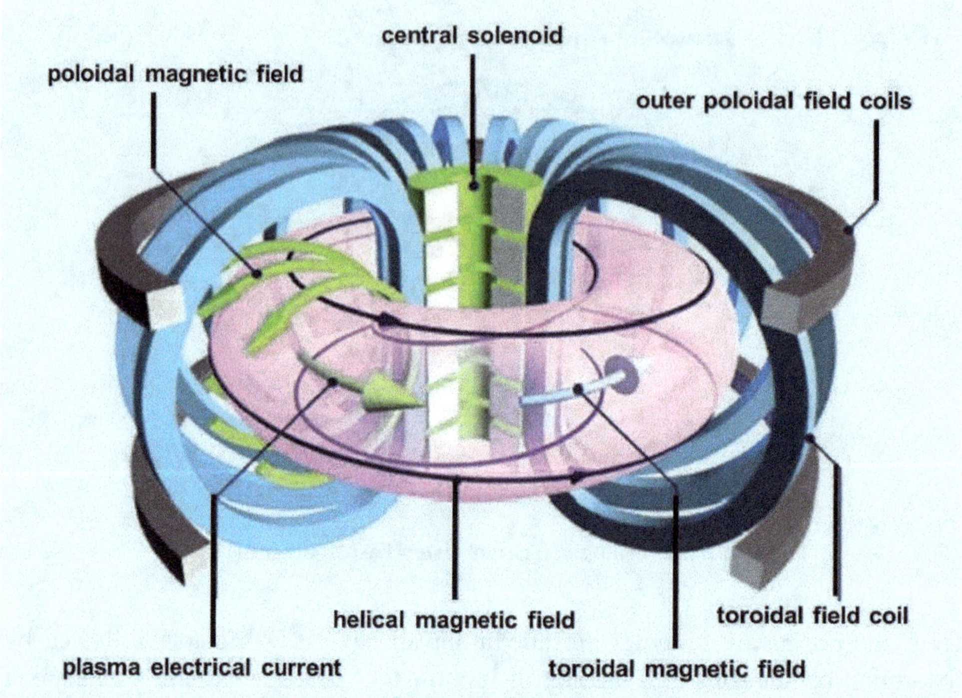

Fig. 12.5: The Principle of Tokamak [189]

It is believed that designs based on tokamaks (toroidal chambers with magnetic coils), first discussed by Russian scientists, are most likely to succeed in harnessing this power. A tokamak consists of a doughnut-shaped vacuum chamber. The toroidal magnetic field helps to hold the plasma. Experiments have been continuing for the last several decades at hundreds of tokamaks installed across the world, including India. These have given valuable insight.

Gaseous hydrogen fuel supplied into the chamber is converted to a plasma under extreme heat and pressure, mimicking the interior of the Sun.

The cloud of the charged particles is controlled by the strong magnetic fields around the vessel as mentioned above. The plasma is kept away from the walls of the vessels and fusion plasma reactions take place in the plasma. Neutrons escape and are absorbed by the wall of of lithium or thourium for breeding of trititum and Uranium-233 respectively.

Fig. 12.6: ITER ("The Way") Collaborators [189]

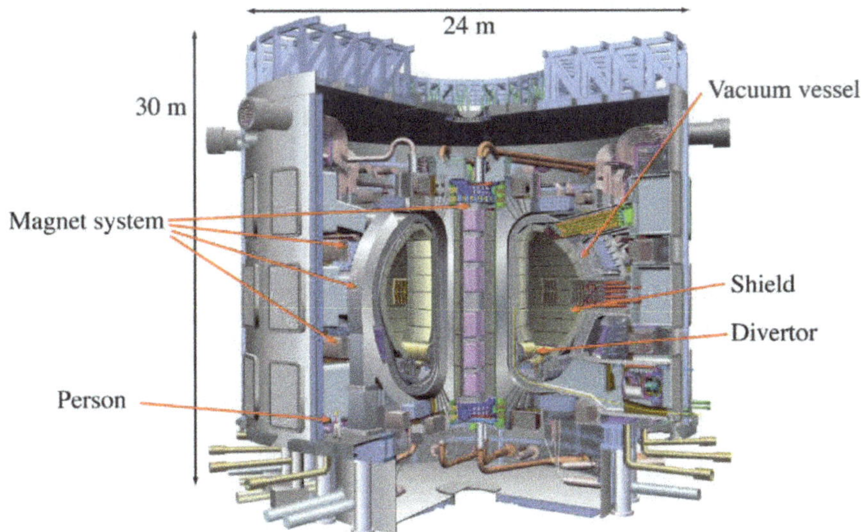

Fig. 12.7: Visualization of ITER [189]

After a great deal of experimentation and theoretical studies, it is now hoped that the ITER (International Thermonuclear Experimental Reactor) collaboration of European Union, India, Japan, South Korea, Russia, China, USA, and many other countries based on the concept of tokamak is likely to start igniting the fusion reaction to produce 500 MW of energy while input energy for heating the plasma would be 50 MW.

The plasma operations are expected to begin in 2025, while the DT fusion is expected in 1935. The total cost of the project is 20 billion dolars. It is hoped that the experience gained will help India set up its own fusion reactor(s) in future. India has remained associated with the project from the beginning and has committed to bear 10% of the cost. It is responsible for providing the Cryostat, In-wall Shielding, Cooling Water Systems, Cryogenic System, Ion-Cyclotron RF Heating System, Diagnostic Neutron Beam System, and Power Supplies.

With ten times the plasma volume of the largest machine operating today, the ITER Tokamak will be a unique experimental tool, capable of longer plasmas and better confinement. The machine has been designed (see [189], for details) specifically to:

1. Produce 500 MW of fusion power
2. Demonstrate the integrated operation of technologies for a fusion power plant
3. Achieve a deuterium-tritium plasma in which the reaction is sustained through internal heating
4. Test tritium breeding
5. Demonstrate the safety characteristics of a fusion device

Several attempts are being made in parallel to ignite the DT pellets by compressing and heating them using laser beams, x-rays, or heavy ions.

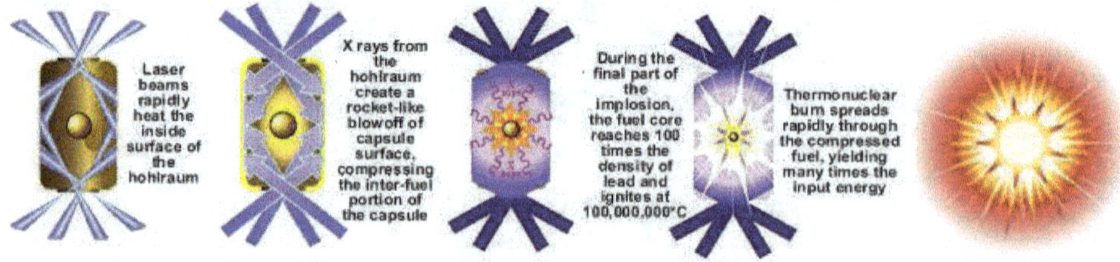

Fig. 12.8: Indirect Drive for Fusion [190]

In one of the methods laser beams heat the inside of a cavity called "hohlraum", which gives out x-rays. These x-rays strike the surface of the pellets which escapes on being heated, compressing the fuel to a very high density and temperature and triggering the fusion reaction. Considerable progress on this front has been made at the National Ignition Facility, at Lawrence Livermore National Facility.

This is called indirect heating or indirect drive fusion.

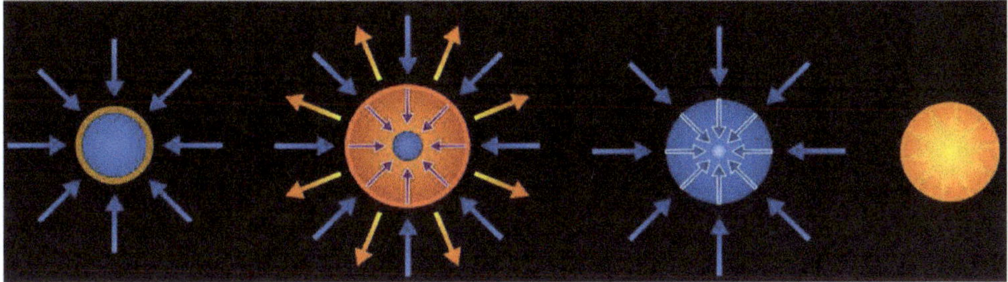

Fig. 12.9: The Direct Drive [191]

In the direct drive, the pellets are subjected to laser beams or heavy ion beams for compressing and heating. This is being attempted at the OMEGA facilty at University of Rochester.

Chapter 13

Accelerator Driven Subcritical System

We have seen that reactor accidents, which resulted in (partial) core meltdown in all the three major nuclear reactor accidents so far, were caused by the problems in the self-sustaining chain reactions which require that the reactor runs in a critical mode. Is it possible to develop new technologies which can displace the conventional nuclear reactors?

Let us discuss the need for this in a little more detail. We have seen that use of Uranium in nuclear reactors produces Plutonium and other minor actinides. Plutonium can be separated and used as fuel. There is a fear that it can be stolen for making a nuclear weapon by a terrorist organization. Similarly, as the spent fuel is highly radioactive, it can be in principle used for making a "dirty bomb" by a terrorist organization, which can be used to disperse radioactivity in populated areas. In order to avoid the potentially destructive usage of Plutonium and other radioactive materials in this manner, several countries have opted to adopt the "once through fuel cycle" so that the spent fuel is first cooled, then vitrified and kept in geological depositories. These may require several thousand years to decay to stable elements or to safe levels. It is also believed that the high level of radioactivity of the spent fuel would be a deterrent against its theft.

The spent fuel is partitioned by several countries, including India, and unspent fuel, Plutonium, and the minor actinides are reused as fuel in "a closed fuel cycle". The fission products can be harvested for useful isotopes. This requires very sophisticated remotely operated liquid chemistry.

The Plutonium-239 thus produced can also be used in Fast Breeder Reactors to produce more Plutonium than they use as well as Uranium-233 from Thorium-232. Uranium-233 can then be used along with Thorium-232 to produce more Uranium-233 than used. This is the 3-stage nuclear programme envisaged by Homi Jehangir Bhabha, which India intends to follow, as discussed earlier. It had its origin in limited resources of Uranium and largest deposits in the world of Thorium in the country. The use of Thorium also gives rise to only a very small concentration of long-lived minor actinides and thus the problem of handling nuclear waste is considerably reduced.

Fast Breeder Reactors require liquid sodium as coolant, which is extremely reactive. There are plans to use molten lead as coolant, which may require still higher temperatures.

We have also noted one issue with conversion of Thorium-232 to Uranium-233. It absorbs a neutron to become Protactinium-233, which gets converted to Uranium-233 after undergoing beta-decay. However, Protactinium-233 can also absorb a neutron and then decay to Uranium-234, whose fission cross-section is very low, and which is useless as a fuel.

The Molten Salt Reactors propose to solve this problem by continuously removing the Protactinium-233 from the core and letting it decay away from the source of neutrons.

All these problems are completely avoided by using an Accelerator Driven Subcritical Systems (ADSS) proposed by Nobel Laureate Carlo Rubia, for which he obtained a patent under the name "Energy Amplifier".

One uses a high energy proton to produce many neutrons by spallation of a heavy nucleus like lead. The neutrons can be used for fission, burning nuclear waste, and conversion of fissile materials like Uranium-238 or Thorium-232 to fertile materials like Plutonium-239 and Uranium-233. The reactor part is always kept at a subcritical level, where the chain reaction at any desired level is sustained by the continuous supply of neutrons.

The entire process is inherently safe as the system can be stopped at any moment by switching off the accelerator.

India has initiated a multi-pronged study of ADSS systems, especially with a view to use its vast Thorium resources. It requires several technological developments. The main technology developments include developing a 30mA, high energy (1 GeV) proton linear accelerator using superconducting Radio Frequency cavities. Construction of lead-bismuth eutectic target for spallation reaction and its cooling mechanism is under progress. It will be tested at 30 MeV, 350 micro-ampere medical cyclotron installed at Calcutta, where a beamline has been specially installed for this.

United States Patent **5,774,514**
Rubbia **June 30, 1998**

Energy amplifier for nuclear energy production driven by a particle beam accelerator

Abstract

A method for producing energy from a nuclear fuel material contained in an enclosure, through a process of breeding of a fissile element from a fertile element of the fuel material via a .beta.-precursor of the fissile element and fission of the fissile element. A high energy particle beam is directed into the enclosure for interacting with heavy nuclei contained in the enclosure so as to produce high energy spallation neutrons. The neutrons thereby produced are multiplied in steady sub-critical conditions by the breeding and fission process. The breeding and fission process is carried out inside the enclosure.

Inventors:	**Rubbia; Carlo** (Geneve, **CH**)
Family ID:	**8213381**
Appl. No.:	**08/632,424**
Filed:	**April 24, 1996**
PCT Filed:	**July 25, 1994**
PCT No.:	**PCT/EP94/02467**
371 Date:	**April 24, 1996**
102(e) Date:	**April 24, 1996**
PCT Pub. No.:	**WO95/12203**
PCT Pub. Date:	**May 04, 1995**

Fig. 13.1: The Patent of Carlo Rubia

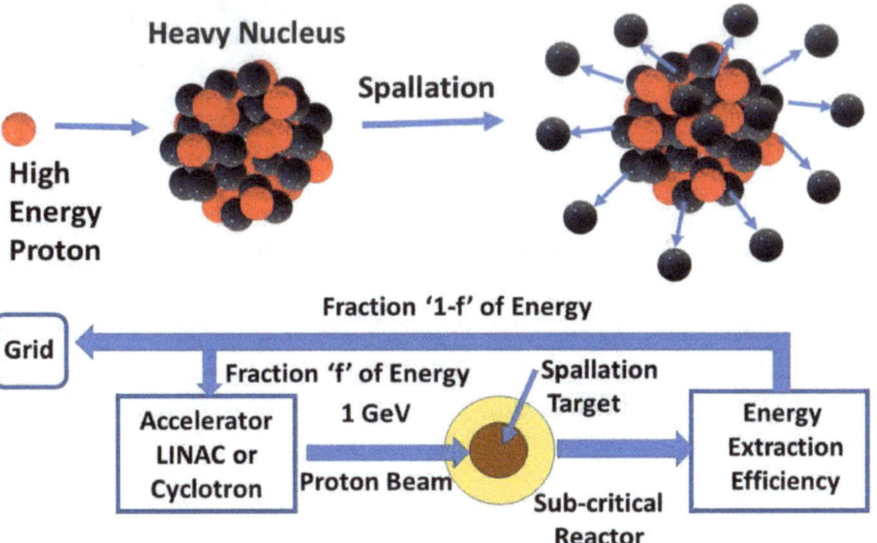

Fig. 13.2: The Principle of Accelerator Driven Subcritical System

The "first of its kind" nuclear reactor based on this concept is under construction in Belgium under the project name MYRRHA (Multi-purpose hYbrid Research Reactor for High-tech Applications).

The reactor will be coupled to a proton accelerator. It will be lead-bismuth cooled. The proton beam would have a current of 4 mA and energy of 600 MeV, operating in continuous mode. The reactor (100 MWt) is expected to operate at a subcriticality level of 0.95. Mixed oxide fuel with enrichment of 30% will be used. It will, additionally, have provisions to use the neutrons for material studies, *etc.*

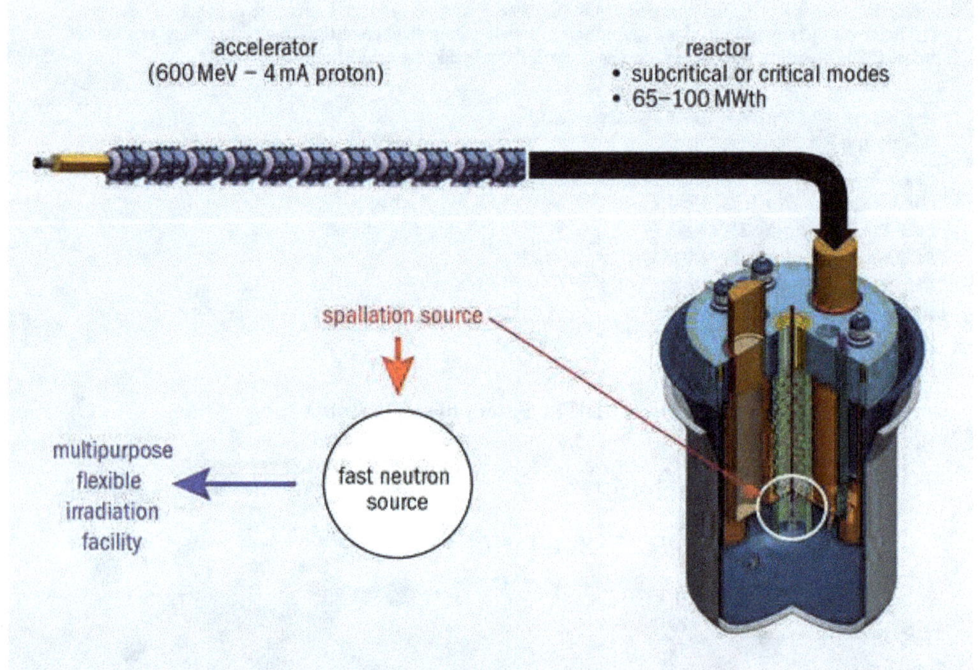

Fig. 13.3: The MYRRAH Experimental ADSS Reactor [192]

France, USA, Japan, and China are also working on these concepts for incineration of nuclear waste and possible utilization of Thorium. If successful, this will open a very safe procedure to incinerate nuclear waste collected over decades.

Chapter 14

Nuclear Safety and Nuclear Waste

Nuclear Safety

Safe running of nuclear reactors, in fact the entire process of the nuclear industry — mining, fuel fabrication, running of reactor, handling of spent fuel and associated components, e.g., cladding, the structure to hold the fuel rod, coolants, moderators, control rods, *etc.*, and transport of radioactive materials — has been studied extensively and International Atomic Energy Agency, Vienna routinely brings out detailed well researched documents providing guidelines for this. India has an independent Atomic Energy Regulatory Board for enforcing these guidelines. It also has a mechanism under which every accident in all nuclear installations, even if very minor, from across the country is reported and analysed in detail and lessons learnt are implemented. It also enforces revised safety guidelines issued by the International Atomic Energy Agency as well as by itself.

This culture of "safety first" in the nuclear industry has ensured that nuclear energy has remained one of the safest sources of energy for mankind. It is interesting to note that the nuclear industry has the lowest mortality rate per 1000 TWh. In contrast, as we have already seen, up to 8 million people die a premature death every year across the world due to ill effects of using fossil fuels in the production of energy.

Now let us look at the experience accumulated by running of nuclear reactors. Only three serious accidents have taken place during this period of about 17000 nuclear reactor years. We shall discuss those as well, in the following. In addition, at present there are about 200 ships (mostly submarines, but also icebreakers and aircraft carriers) powered by nuclear reactors. Many of the nuclear submarines have now been retired. India commissioned its first nuclear submarine, Arihant in 2016. Taken together, these amount to additional 13,000 nuclear reactor years of operation. The safety record is exemplary.

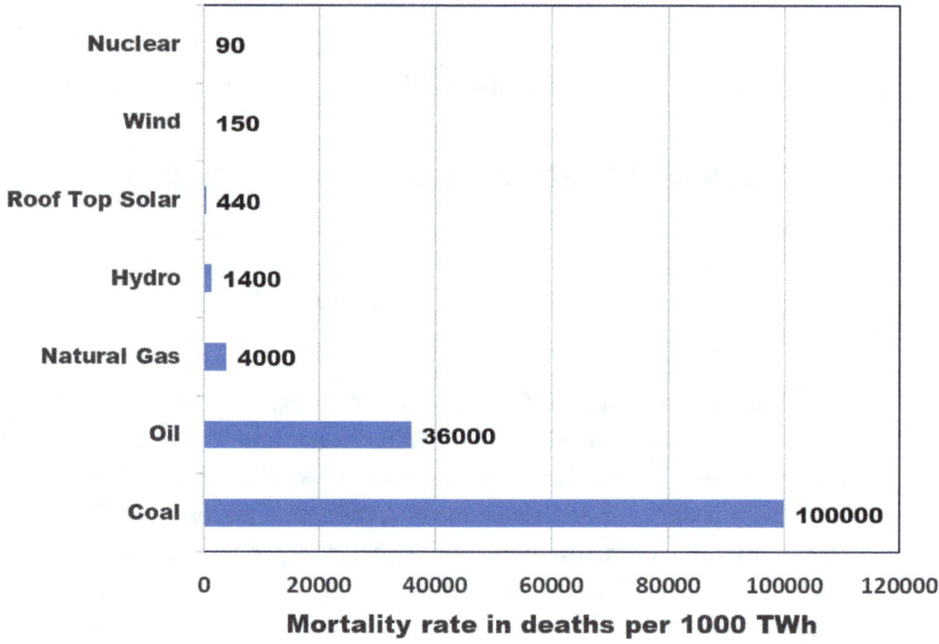

Fig. 14.1: A Comparison of Mortality Rate for Various Sources of Energy [193]

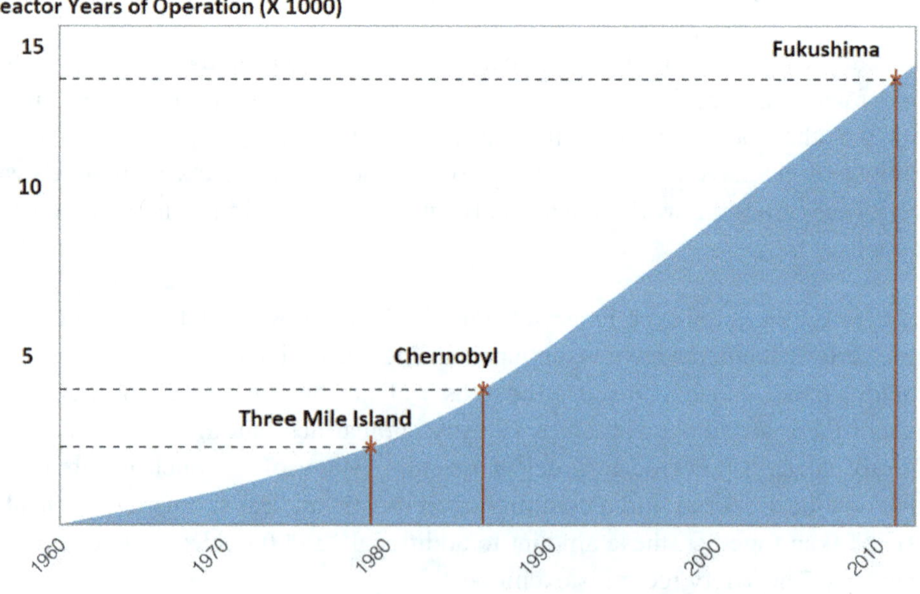

Fig. 14.2: Cumulative Years of Reactor Operation [194]

Defence in Depth

The safety in nuclear reactors utilizes the concept of several physical barriers ("defence in depth") between the radioactive reactor core and the environment:

1. The first barrier is the fuel itself. We have seen that it is in the form of solid ceramic pellets and radioactive fission products remain inside the pellet as the fuel is burnt.

2. The second barrier is fuel rod. The fuel pellets are packed inside zirconium alloy tubes. These are hermetically sealed and designed to contain radioactive material under the extreme conditions inside the reactor core.

Fig. 14.3: Reactor Vessel, India

3. The third barrier is a hermetically sealed primary circuit pressure system with the reactor pressure vessel preventing release of coolant with radioactive substances into the environment at all anticipated temperatures and pressures. The vessel is made of special fine-grained low alloy ferritic steel, well suited for welding and with a high toughness while showing low porosity under neutron irradiation. The inside is lined with austenitic

steel cladding to protect against corrosion. It is normally 35 cm thick or still thicker.

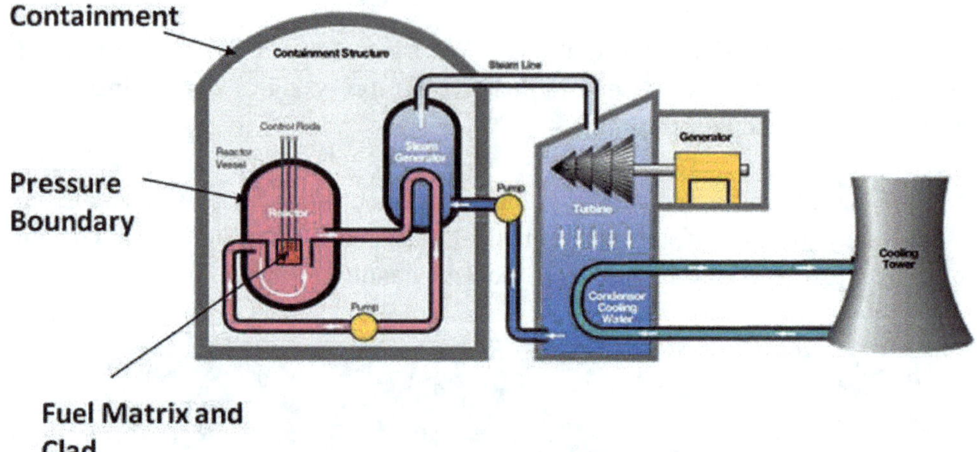

Fig. 14.4: Containment Building for Pressurised Water Reactors [195]

4. The fourth barrier is the containment — 1.5-metre-thick steel-concrete envelope, retaining radioactive substances that could be released through potential damage of the first three barriers. These are mostly made of double walls to withstand missile attacks, severe earthquakes, and flooding, *etc.*

In addition, many of the newer reactors are being designed and constructed with a core catcher, to be installed just below the reactor vessel. It is a device provided to catch the molten core material of a nuclear reactor in case of a nuclear meltdown and prevent it from escaping the containment building. It is made from a special concrete ceramic to prevent material from trickling through. It is also provided with a cooling mechanism to cool down the core material.

Fig. 14.5: Installation of Core Catcher at Rooppur Nuclear Power Plant, Bangladesh

In India, a nuclear power plant site is generally selected in a relatively low-population zone. This has the basic objective of limiting the number of people who receive a dose of radiation under normal conditions or during an accident.

This is achieved by dividing the area surrounding nuclear power plants into following zones.

a. Exclusion zone: An exclusion zone of 1.5 km radius around the plant is established, which is under the exclusive control of the operating organization, and no public habitation is permitted in the area.

b. Sterilised Zone: Efforts are made to establish a sterilized zone up to a 5-km radius around the plant. This is the annulus around the exclusion zone, which has the potential for extensive contamination in case of a severe accident. Development activities within this area are controlled.

c. Emergency Planning Zone: This is the zone defined around the plant up to a 16-km radius. It provides for the basic geographical framework for decision making while implementing measures as a part of graded response in the event of an off-site emergency.

These considerations are based on a premise that radiation received by public and radiation workers should not exceed the prescribed limit considered extremely safe because of their activities. Radiation dose is measured in milli-Sievert (mSv) and radiation exposure is measured in milli-Gray (mGy). The absorbed dose is measured in Gray (Gy), which represents the amount of radiation required to deposit one joule of energy in one kilogram of any type of matter. The Sievert (Sv) is unit of equivalent radiation dose. One Sievert is the amount of radiation necessary to produce the same effect on living tissue as one Gray of high-penetration x-rays or gamma rays. Quantities that are measured in Sieverts are designed to represent the biological effects of ionizing radiation.

According to the Report on "Evaluation of Radiation Exposure Guidance for Military Operations (1997)" by J. Christopher Johnson and Susan Thaul, "A whole-body exposure of 500 mGy leads to radiation sickness. A dose of 3000 mGy leads to acute gastrointestinal syndrome, which can be fatal without major medical intervention. A dose of 5000 mGy is a lethal dose which can kill 50% of the people in thirty days, even with treatment. These are acute thresholds: the same dose fractionated (distributed) over a series of exposures or over a longer time may produce less injury as the body has a chance to repair damage. Radiation doses that exceed a minimum (threshold) level can cause undesirable effects such as depression of the blood cell-forming process (threshold dose = 500 mSv) or cataract (threshold dose = 5000 mSv). The scope and severity of these effects increases as the dose increases. Radiation can also cause an increase in incidence, but not the severity, of malignant disease (e.g., cancer). For radiation protection purposes, it is assumed that any dose above zero can increase the risk of radiation-induced cancer. Epidemiologic studies have found that the estimated lifetime risk of dying from cancer is greater by 0.004% per mSv dose of radiation to the whole body."

Table 14.1: Dose Limitations [196]

Part of Body	Radiation Workers	Public
Whole body (Effective dose)	20 mSv/year averaged over 5 consecutive years	1 mSv/year
Lens of eyes (Equivalent dose)	150 mSv in a year	15 mSv/year
Skin (Equivalent dose)	500 mSv in a year	50 mSv/year
Hand and feet (Equivalent dose)	500 mSv in a year	-

For pregnant radiation workers, after declaration of pregnancy, the dose on the embryo/foetus should not exceed 1 mSv.

The recommended values for limits of safe dose of radiation for public and radiation workers, are extremely safe, indeed.

To put these numbers and recommendations in a proper perspective, let us look at the dose of radiation we receive during our normal routine activities, from natural sources.

Table 14.2: Radiation Dose from Natural Sources

Source	Dose
Cosmic Rays	0.29 mSv/year
Terrestrial	0.57 mSv/year
Radon (Concrete, Rocks)	1.76 mSv/year
Cosmogenic (C-14 *etc.*)	0.01 mSv/year
Food (Milk, Banana)	0.32 mSv/year
Total	2.98 mSv/year

We add that there are several places in the world where the natural background radiation is much higher and where people have been living for generations without any ill effect. Thus, for example the Ramsar town in northern Iran, people receive an annual dose of 260 mSv/year. The dose of background radiation received by people, from monazite sand beaches in Kerala ranges from 4 mSv/year to 70 mSv/year. Yangjiang in China and Guarapari in Brazil are also known for high background radiations where people have been living for generations.

Table 14.3: Radiation Dose from Various Activities

Activity	**Dose**
Air Travel (1 hour/year)	0.005 mSv/year
X-ray (1/year)	0.2 mSv/year
CT scan (1/year)	6.6 mSv/year
Staying Close to Nuclear Power Plant	
Tarapur	0.007 mSv/year
Rawatbhata	0.015 mSv/year
Narora	0.0005 mSv/year
Kakrapar	0.001 mSv/year
Kaiga	0.001 mSv/year

The discussion so far, suggests that if the necessary safeguards are properly implemented nuclear reactors are very safe.

However, no discussion on nuclear safety is complete without a discussion of three nuclear accidents; The Three Mile Island, Chernobyl, and Fukushima, which slowed down the progress of installation of nuclear reactors considerably and have led to a feeling of fear psychosis in the minds of people about nuclear energy.

The Three Mile Island Accident

The accident took place at the Three Mile Island Nuclear Power Plant in USA, in which a cooling mechanism malfunction caused a part of the core to melt. Some radioactive gas was released after the accident, but not enough to cause any dose above background level to residents, nearby. There were no injuries or adverse health effects, either.

A vast body of literature, with a critical analysis of the sequence of events leading to the accident and its consequences for the construction of nuclear reactors across the world is available. The account given here closely follows the details provided by the World Nuclear Association.

The accident involved the Three Mile Island's second unit (TMI-2), a pressurised water reactor, with a capacity of 906 MWe. The accident took place around 4 AM on 28 March 1979, when the reactor was operating at 97% power.

"It involved a relatively minor malfunction in the secondary cooling circuit which caused the temperature in the primary coolant to rise. This in turn caused the reactor to shut down automatically. Shut down took about one second. At this point a relief valve failed to close, but instrumentation did not reveal the fact, and so much of the primary coolant drained away that the residual decay heat in the reactor core was not removed. The core suffered severe damage as a result."

There was a build-up of radioactive gases from the reactor cooling system, which were transferred to waste gas decay tanks. The compressors leaked, and some radioactive gases were released to the environment. However, as these went through charcoal and high-efficiency particulate air filter, most of the radionuclides except for noble gases were stopped. As these were chemically and biologically inert and additionally had short half-lives, they did not pose a health hazard.

A critical scrutiny has determined that deficient control room instrumentation and inadequate emergency response training were the root causes of the accident. The clean-up of the damaged nuclear system was performed without any mishap and took about 12 years at a cost of approximately 973 million US dollars.

This accident had serious consequences for the future construction of nuclear reactors across the world and in the United States of America, where no new reactors were built for thirty years.

Chernobyl

The Chernobyl accident took place in April 1986. It is attributed to a flawed reactor design that was operated with inadequately trained personnel. It is also believed that the accident was a direct consequence of Cold War isolation and the resulting lack of safety culture.

The accident has been widely and critically analysed and a vast literature is available for it. We shall discuss the main points here provided by World Nuclear Association and IAEA.

The RBMK-1000, reactor involved in the accident used graphite as moderator and a lightly enriched Uranium (2% U-235) as fuel. It had a power output of 3200 MW and an electrical output of 1000 MW. It was a boiling water reactor, where water pumped at the bottom of the fuel rods boiled as it moved up, working as a coolant, and providing steam to run the turbines. The pressure tubes contained the zircaloy clad fuel rods around which the water flowed.

The biggest flaw in the reactor design was a "positive void coefficient". In these reactors the graphite provided the moderation (or slowing down) of neutrons as stated above. The (cooling) water absorbs neutrons and lowers the reactor activity. However, when voids are created by formation of steam, this removal of neutrons due to absorption is reduced, as steam has a much lower density than water. Thus, unabsorbed moderated neutrons increase the activity and further heat up the reactor. This leads to more water being converted to steam and creating further void and increasing the activity of the reactor.

The reactivity is brought under control using control rods, which can quickly absorb neutrons. However, there were two problems. The control rod movement mechanism was antiquated. The control rods also had a design flaw of being tipped with graphite instead of neutron poison (cadmium or boron), which displaced water as it moved, adding to the activity, instead of absorbing the neutrons and reducing the activity.

Finally, the reactor did not have a "containment building", which could have contained the released radioactivity.

"On 25 April, prior to a routine shutdown, the reactor crew at Chernobyl-4 began preparing for a test to determine how long turbines would spin and supply power to the main circulating pumps following a loss of main electrical power supply…. A series of operator actions, including the disabling of automatic shutdown mechanisms, preceded the attempted test early on 26 April. By the time that the operator moved to shut down the reactor, the reactor was in an extremely unstable condition."

The positive void coefficient and the peculiarities of the control rods mentioned above caused a dramatic power surge as they were inserted into the reactor. It is believed that the rise of neutron flux burnt away Xe-135, which acts as a poison, due to its very large neutron absorption cross-section of 2.6 million barns. This also added to the surge in power and instability.

"The interaction of very hot fuel with the cooling water led to fuel fragmentation along with rapid steam production and an increase in pressure... The overpressure caused the 1000-ton cover plate of the reactor to become partially detached, rupturing the fuel channels, and jamming all the control rods, which by that time were only halfway down. Intense steam generation then spread throughout the whole core ... causing a steam explosion and releasing fission products to the atmosphere. About two to three seconds later, a second explosion threw out fragments from the fuel channels and hot graphite." The second explosion was most possibly caused by production of hydrogen due to reaction of zircaloy with water at high temperature... Two workers died because of these explosions. The graphite and fuel became incandescent and started several fires, causing the main release of radioactivity into the environment. A total of about 14 EBq (14×10^{18} Bq) of radioactivity was released, over half of it being from biologically-inert noble gases."

We add that 1 Bq (Becquerel) of radioactivity corresponds to 1 nuclear decay per second of a radioactive material.

Two Chernobyl plant workers died on the night of the accident, and a further 28 people died within a few weeks because of acute radiation poisoning. Nobody offsite suffered from acute radiation effects although significant, but uncertain, fraction of the thyroid cancers (around 11 at the time of writing this) diagnosed since the accident, in patients who were children at the time, are likely to be due to intake of radioactive iodine fallout. Furthermore, large areas of Belarus, Ukraine, Russia, and beyond were contaminated in varying degrees.

Initially about 116,000 people and then additional 220,000 people were resettled outside 4300 square kilometres around the plant. Some of them have now returned to live in their old dwellings even though the background radiation is still somewhat high.

Conifers in about 10 square kilometres of forest close to the plant were killed by the high radiation levels, but regeneration got underway from the following year. The net environmental effect of the accident has been much greater biodiversity and abundance of species. The exclusion zone has become a unique sanctuary for wildlife, though the extent of their radiation sickness is not known.

The initial days after the accident were a time of extreme panic, trauma, and confusion. However, the continued monitoring of the people for over two decades has not revealed any adverse effect on their health.

Several RBMK reactors are still operating in former USSR. Following the accident, these have been modified to provide a negative void coefficient along with several new safety features and modern control systems.

The nuclear reactor scientists and engineers of former USSR are in much closer contact with IAEA and other countries building and operating nuclear reactors, leading to valuable scrutiny of their efforts by experts from across the world.

Most of the new reactors under construction are now being built with core catcher.

Fukushima

There is a considerable amount of literature on the accident. The account given below is mostly based on the reports of World Nuclear Association.

Fukushima Daiichi Nuclear Power Plant (now disabled) is in the towns of Okuma and Futaba in Fukushima, Japan. It was commissioned in 1971 and had 6 boiling water reactors. A neighbouring site, Fukushima Daini Nuclear Power Plant about 12 kilometres away has 4 reactors.

The Daiichi site was originally a bluff (a cliff) 35 metres above sea level, but its height was lowered by 25 metres during construction, so that the reactors could be built on a solid bedrock and the cost of pumping cooling water to the greater height could be avoided.

At 2.46 PM on Friday, 11 March 2011, The Great East Japan Earthquake of magnitude 9.0 hit the region and the accompanying 14 to 15-metre-high tsunami caused a considerable damage.

"The earthquake was centred 130 km offshore the city of Sendai in Miyagi prefecture on the eastern coast of Honshu Island (the main part of Japan), and was a rare and complex double quake of a severe duration of about 3 minutes. An area of the seafloor extending 650 km north-south moved typically 10–20 metres horizontally. Japan moved a few metres east and the local coastline subsided half a metre. The tsunami inundated about 560 sq. km and resulted in a human death toll of about 19,000 and much damage to coastal ports and towns, with over a million buildings destroyed or partly collapsed."

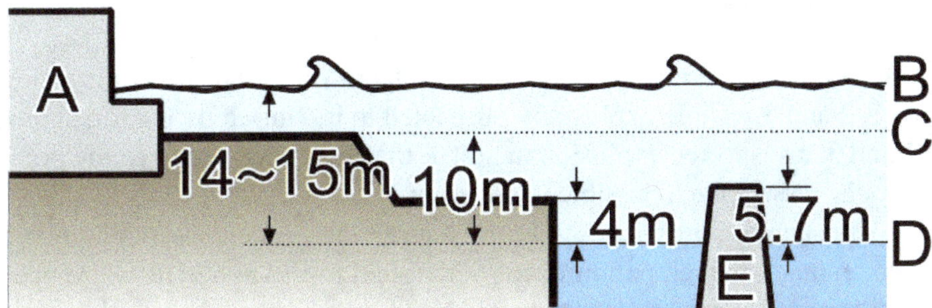

A: Power station buildings, B: Peak height of tsunami, C: Ground level of site, D: Average sea level, E: Seawall to block waves

Fig. 14.6: The Height of the Tsunami that struck the Fukushima Daiichi Station Approximately 50 minutes after the Earthquake [197]

Units 1, 2, and 3 at Daiichi were operating before the earthquake, while the Units 4, 5, and 6 were already shut down for maintenance. These 3 units along with other 8 reactors in the neighbourhood, including the 4 units at Daini, shutdown in response to the earthquake. Subsequent checks revealed that all the reactors withstood the earthquake. The decay heat of the reactors was being removed by emergency diesel generators when the first tsunami hit the site at 3:42 PM followed by another one 8 minutes later. These reached a height of 15 metres and inundated the site completely, thereby disabling the emergency power supply required to cool the reactors and the spent fuel. The first earthquake was followed by hundreds of aftershocks and 3 more earthquakes of magnitude around 7. These, however, did not cause any structural damage to the reactors.

The tsunamis "submerged and damaged the seawater pumps for both the main condenser circuits and the auxiliary cooling circuits, notably the residual (decay) heat removal cooling system as mentioned above. They also drowned the diesel generators and inundated the electrical switchgear and batteries, all located in the basements of the turbine buildings (the one surviving air-cooled generator was serving units 5 & 6). Thus, there was a station blackout, and the reactors were isolated from their ultimate heat sink. The tsunamis also damaged and obstructed roads, making outside access difficult."

The emergency power supply at the Daini plant had also been damaged, but the height of the tsunami was only 9 metres there and extraordinary effort by the staff succeeded in establishing an emergency cooling system for them. The reactors at other sites were protected from tsunami.

Over the following three weeks there was evidence of partial nuclear meltdowns in units 1, 2 and 3: visible explosions, suspected to be caused by hydrogen gas, in units 1 and 3; a suspected explosion in unit 2, that may have damaged the primary containment vessel; and a possible uncovering of the spent fuel pools in Units 1, 3 and 4. As the cooling provided by the only surviving generator serving units 5 & 6 was not enough, the temperature was also found to rise and vents were opened in the ceilings of those buildings to avoid build-up of gases, especially hydrogen and possible explosion.

The release of radiation from Units 1-4 forced the evacuation of 83,000 residents from the neighbourhood. The high radioactive release over 4 days amounted to 940×10^{15} Bq.

There have been no deaths or cases of radiation sickness from the accident.

The clean-up of the plant may take about 40 years of sustained work and is expected to cost 75–660 billion dollars.

There had been repeated warnings based on more accurate simulations and study of other earthquakes in the Pacific region, that major earthquake in the area could give rise to tsunamis reaching a height of 15.7 metres. This required that the emergency power supply for removal of decay heat should have been moved to a greater height. The warnings and advisories were ignored.

The tsunami countermeasures could also have been reviewed in accordance with IAEA guidelines which required taking into account high tsunami levels, but the Nuclear and Industrial Safety Agency of Japan continued to allow the Fukushima plant to operate without sufficient countermeasures, such as moving the backup generators up the hill, sealing the lower part of the buildings, and having some back-up for seawater pumps, despite clear warnings.

The Fukushima accident, along with those at Three Mile Island and Chernobyl, gave a strong voice to the anti-nuclear activists and they have succeeded in getting their governments to shut down, go slow, or delay construction of nuclear plants, in several countries, including India.

Before closing, we add that we have discussed the Nuclear Safety in detail to emphasize the fact that nuclear scientists and engineers are very keen on it and that the reactors are continuously incorporating latest safety aspects.

Nuclear Waste

We have seen that uranium and thorium resources of the world can meet the energy needs of the entire world for several thousand years. Nuclear reactors provide continuous power in plenty and have a life of up to one hundred years.

Yet nuclear power has met with resistance basically on two counts; it generates nuclear waste which is radioactive, and the nuclear reactors have had some severe accidents. There are also concerns of diverting plutonium for making a bomb or of diversion of nuclear waste to make a "dirty bomb" which can spread radioactivity. The other concern which has not been addressed in enough detail is the cost of decommissioning a reactor after it has lived its useful life. Countries like Germany, which have decided to close their nuclear reactor facilities prematurely have just started dealing with this problem. The cleaning up of sites involving nuclear accidents (see e.g., Chernobyl and Fukushima) can take considerable time and involve considerable expenditure running into hundreds of billions of dollars.

Nuclear waste is the spent nuclear fuel in which the fissile material has depleted to a level where it cannot sustain a chain reaction. We have seen that the fission of U-235 typically produces a distribution of fission products which have peaks around mass numbers of 95 and 137. These are mostly highly radioactive and decay over a few hundred years to stable nuclei.

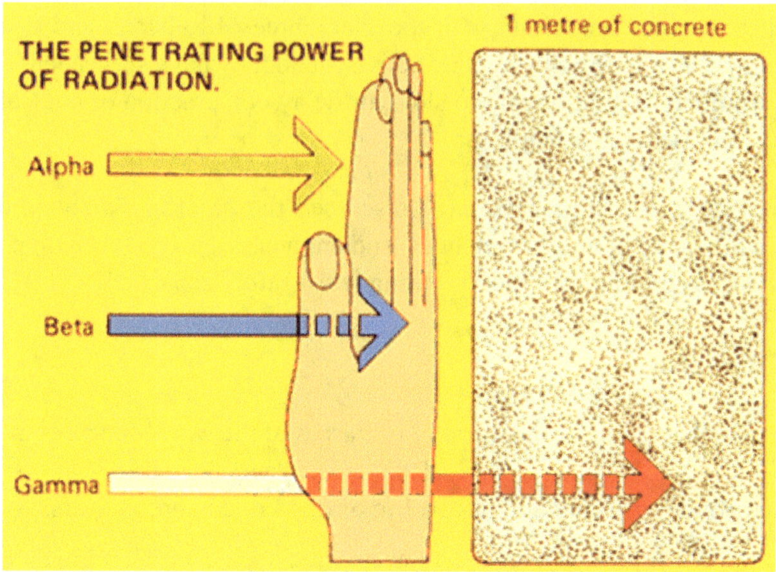

Fig. 14.7: Alpha, Beta, and Gamma Radiations

It may be added that radioactive nuclei give out alpha (nucleus of He-4), beta (electron) and gamma radiations. The highly excited fission products will also give out (delayed) neutrons on a timescale of 10^{-14} seconds, as mentioned earlier.

Alpha particles are easily stopped even by thinnest of materials like a thin sheet of paper. Beta radiations are ionizing and can penetrate human body and can be stopped by a thin concrete or a thin sheet of metal. Gamma radiations need thick concrete shielding, which can also reduce the number of neutrons which may leak-out. This must be remembered while making water pools and casks for storing the radioactive waste (see later).

The absorption of neutrons by U-238 and its beta decays leads to the production of plutonium and other minor (transuranic) minor actinides nuclei like neptunium, americium, and curium — which are also radioactive, but mostly with decay chains lasting over several tens or hundreds of thousands of years.

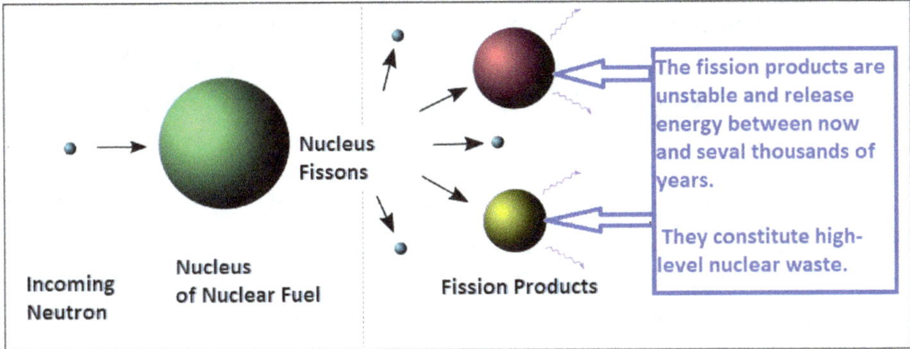

Fig. 14.8: Generation of Nuclear Waste [198]

Reactors using thorium (which will be converted to uranium-233) do not produce long-lived plutonium or minor actinides simply because it would require absorption of six or more neutrons by the same nucleus accompanied by beta decays.

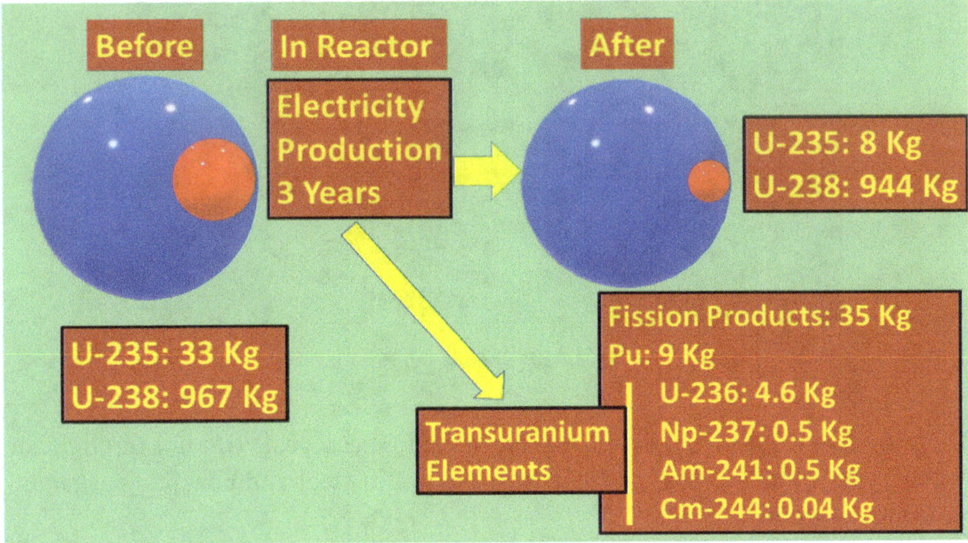

Fig. 14.9: Spent Fuel from a Light Water Reactor after using 1 ton of Uranium Enriched to 3.3% for Three Years [199]

Fast Breeder Reactors burn plutonium and can also burn minor actinides and eliminate long lived radioactive products. Fast neutrons are also known to be very efficient for burning these.

Let us get back and consider a light water reactor using uranium enriched to 3.3%. Consider 1000 kilograms of enriched uranium. It would mean about 33 kilograms of U-235 and 967 kilograms of U-238. The composition of the spent fuel can be seen in the accompanying illustration.

The spent fuel from Heavy Water Reactors has about 0.25% U-235, 0.35% Pu, 3–5% percent fission products and a small percentage of minor actinides. The rest is U-238. This waste as, mentioned above, is very hot and highly radioactive. It is usually kept in a water pool which is ten to fifteen meters deep for about 20 years or more till it is safe enough for remote handling. The water needs to be cooled and circulated. USA alone has accumulated 70,000 tons of spent fuel as waste and produces 2000 tons/year.

Fig. 14.10: Spent Fuel Pool

There are two ways to handle this waste. In the first case, called "once through" of "open cycle", the "cooled" nuclear waste is put in steel and concrete casks and stored underground.

There are plans to vitrify the nuclear waste and store it in deep and stable geological formations. However, so far only Finland has finalized such a site, as there have been opposition to the storage of waste by people due to the fear of leaching of radioactivity and contamination of sources of water, *etc.*

Fig. 14.11: Dry Cask Storage of Spent Fuel [200]

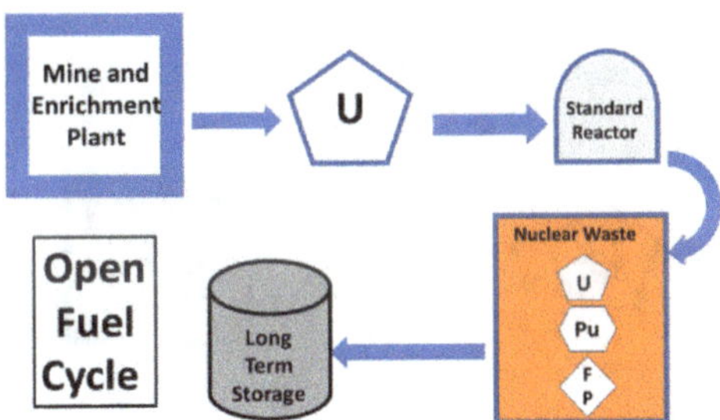

Fig. 14.12: Open or Once-Through Fuel Cycle [201]

It is believed that this route to disposal of nuclear waste has too dilute a concentration of plutonium for stealing. The radioactivity of the waste products is also believed to act as an effective deterrent against its theft. USA and several countries in Europe have resorted to this procedure for waste disposal. However, this route essentially postpones the problem of handling nuclear waste to future generations. The waste thus stored is likely to remain radioactive for several thousand years. In addition, one also loses valuable fuel.

We have seen that the spent nuclear fuel of uranium based nuclear reactors has a large amount of (depleted) uranium, plutonium, and minor actinides which can be again used as fuels. The fission products include many useful isotopes like Caesium-137, Strontium-90, Ruthenium-106, *etc.*, which can be harvested. This is called "Closed Fuel Cycle". It is practiced by India.

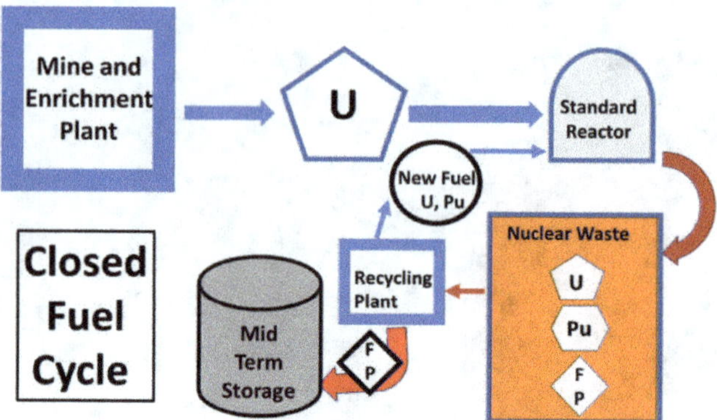

Fig. 14.13: Closed Fuel Cycle [201]

We add that, as discussed earlier, reactors using heavy water as moderator will not need enrichment of uranium fuel. We have seen that breeder reactors produce more fuel than they consume. Their fuel cycle must be necessarily closed and proceed as follows.

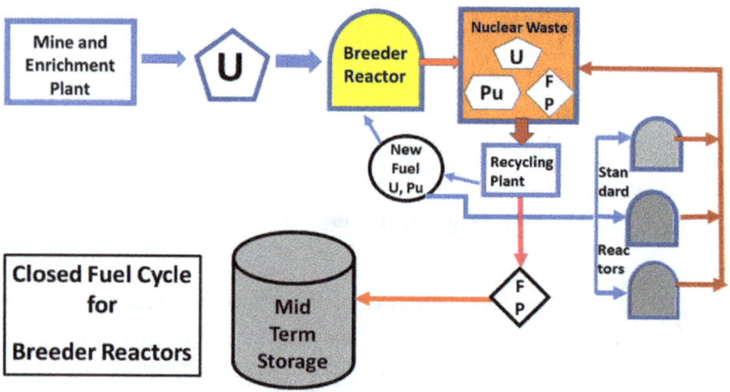

Fig. 14.14: Closed Fuel Cycle for Breeder Reactors [201]

Several countries like India, Japan, Germany, and China have accumulated a vast experience in reprocessing the nuclear fuel. There are ongoing efforts to make batteries using some radioactive isotopes in nuclear waste. There is also ongoing research to use strong laser beams for the incineration of the nuclear waste. The laser beams fired at a gold target produce high energy photons, which in turn transmute the long-lived radio-active nuclei to short-lived or stable nuclei via a (gamma, n) nuclear reaction.

Chapter 15

Healthcare and other Applications of Nuclear Radiations

In the preceding pages, we have given indications of several applications of nuclear radiations. Scientists have studied nuclear radiations since 1890s. This has led to a wide variety of its uses. Now, we use radiation in medicine, academics, and industry. It has become a valuable tool of investigation in such varied areas as agriculture, archaeology, geology (including mining), law enforcement and space exploration, *etc.*

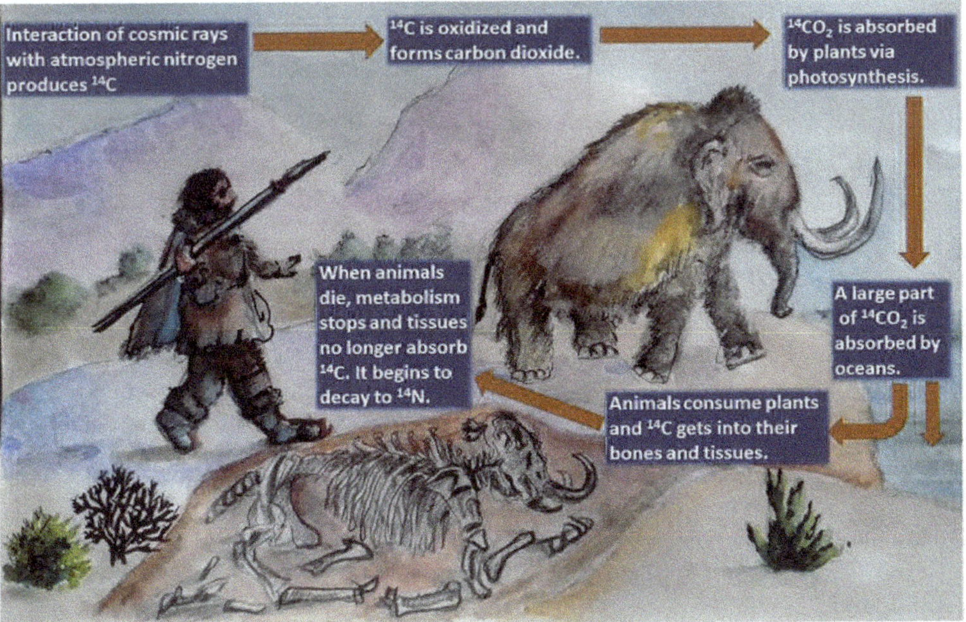

Fig. 15.1: Carbon-14 Dating [202]

The use of carbon dating in archaeology has been very valuable for a reasonably accurate determination of the age of fossils and artifacts of historical importance.

It works on a very simple principle. Carbon-14, a radioactive isotope of carbon is continuously produced and decays in the atmosphere due to the action of cosmic radiations. Its concentration in the atmosphere reaches an equilibrium. Carbon dioxide is absorbed by plants and is converted to food for plants. These plants are in turn eaten by animals. As long as the animal lives, the carbon is continuously replenished, and an equilibrium is maintained. Once the animal dies, this process is stopped, and carbon-14 starts decaying. The half-life of carbon-14 is 5730 years. The remaining carbon-14 then gives us the time when the animal died.

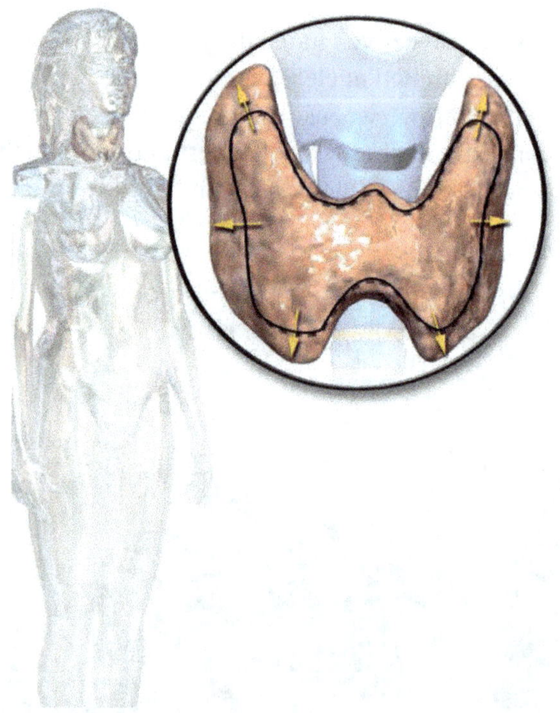

Fig. 15.2: I-131 is Used to Treat Hyperthyroidism [203]

Doctors have developed a variety of nuclear materials and procedures to diagnose, monitor, and treat a wide assortment of metabolic processes and medical conditions in humans. Use of radiation has saved thousands of lives through the detection and treatment of conditions ranging from hyperthyroidism to bone cancer.

Of-course, the use of X-rays to detect damaged bones, teeth, and several kinds of infections, e.g., in the lungs, is well known. It has also been used for a long time to kill cancerous tissues, reduce the size of a tumour, or to reduce pain. X-ray along with computers have been used to develop computerized axial tomography (CAT) or computed tomography (CT) scanners, which provide doctors with coloured images which show the shapes and details of internal organs. This is very valuable in locating and identifying tumours, their anomalies and their physiological or functional organ problems.

Nuclear medicine procedures are adopted across the world in large numbers, every day. In these procedures, doctors administer slightly radioactive substances (e.g., Tc-99m) to patients, which are attracted to certain internal organs for imaging and functional studies of the brain, myocardium, thyroid, lungs, liver, gallbladder, kidneys, skeleton, blood, and tumours.

Positron Emission Tomography (PET) uses a positron emitting isotope (e.g., Flourine-18 or Carbon-11), which is administered to the patient. The isotope emits a positron, which is annihilated by an electron, leading to back-to-back gamma rays of about 511 eV, which are detected by detectors along a ring surrounding the patient.

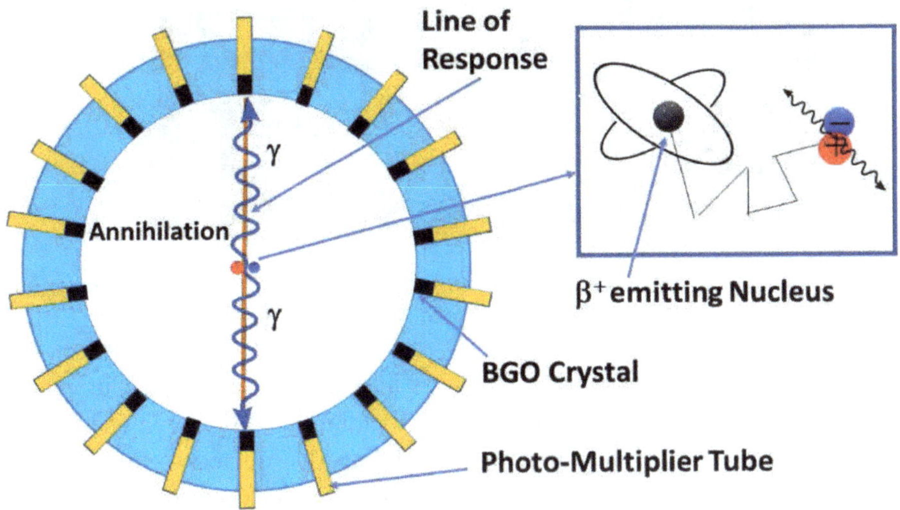

Fig. 15.3: Principle of Positron Emission Tomography [204]

PET is valuable for the study of the dynamics of substance transport, diffusion, migration, and accumulation process with a time resolution down to 1 second.

A single photon emission (from an isotope of Iodine-123, Iodine-125, Indium-111, *etc.*) computed tomography (SPECT) scan is an imaging test that shows how blood flows to tissues and organs. It is used to diagnose seizures, stroke, stress fractures, infections, and tumours in the spine.

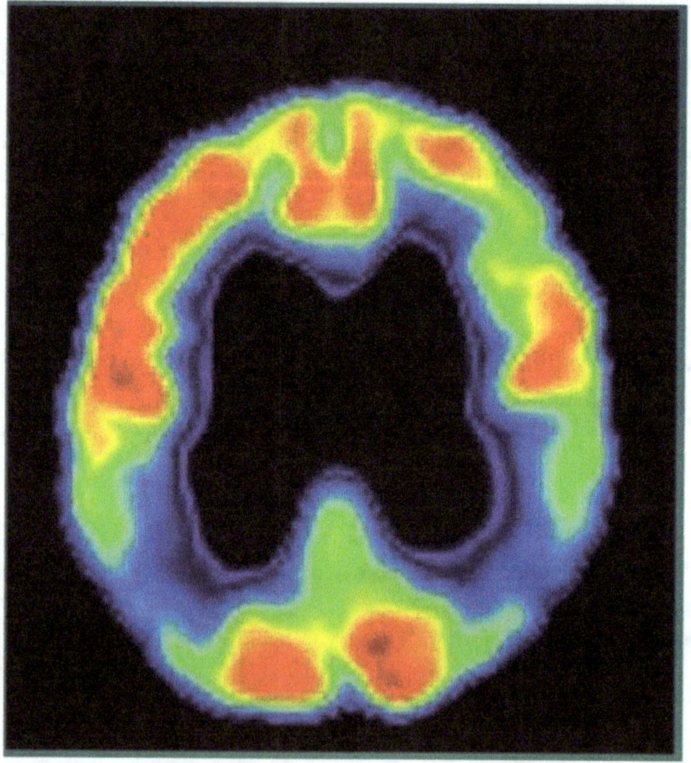

Fig. 15.4: A 99mTc-ECD (ethylene-cysteine-dimer) Brain SPECT Image demonstrating Bilateral Temporoparietal Hypoperfusion Typical of Alzheimer's Disease [205]

The germ-killing property of radioactivity was identified by Madame Curie, right in the early days and X-rays as well as nuclear radiations have been used to kill germs without harming the substance that is being disinfected — and without making it radioactive. When treated in this manner, foods take much longer to spoil, and medical equipment (such as bandages, hypodermic syringes, and surgical instruments) are sterilized without being exposed to toxic chemicals or extreme heat. This is now extensively used to increase shelf-life of potatoes and food items.

Municipal sludge is often disposed in unorganized manner resulting in environmental pollution and spread of diseases. The sludge produced carries a heavy microbiological load. Sludge also contains worms, ova, viruses, helminthic, weeds, *etc.* It also contains toxic heavy metals and organic pollutants like pesticides, poly-aromatic hydrocarbons, drugs, and other persistent pollutants. Sludge is a rich source of many macro (nitrogen, phosphorous, potassium), and micronutrients (zinc, iron, copper, manganese) and organic carbon essential for soil.

Fig. 15.5: 100 Tons Per Day Sludge Treatment Plant in Ahmedabad

Radiation (e.g., from Co-60) can be used to kill pathogens in municipal sludge, which can then be used as fertilizer. Bhabha Atomic Research Centre has set up a plant loaded with 150 kCi of Co-60 in collaboration with Amdavad Municipal Corporation, Ahmedabad. This "Sewage Sludge Hygienisation Plant" at Shahwadi, Ahmedabad was inaugurated in February 2019 and is in continuous operation since then.

Another liquid sludge irradiator: Sludge Hygienisation Research Irradiator (SHRI) is operating at Vadodara for radiation treatment of raw sludge containing 3–4% solids since last 30 years. There is an urgent need to install these plants at all large cities of the country and convert the sludge to a valuable fertilizer, instead of disposing this toxic waste.

The sewage water can also be treated with 1 MeV high current electron beam to kill bacteria and pathogens, so that the water can be safely used for irrigation. Some plants have already been set up across the country using this phenomenon.

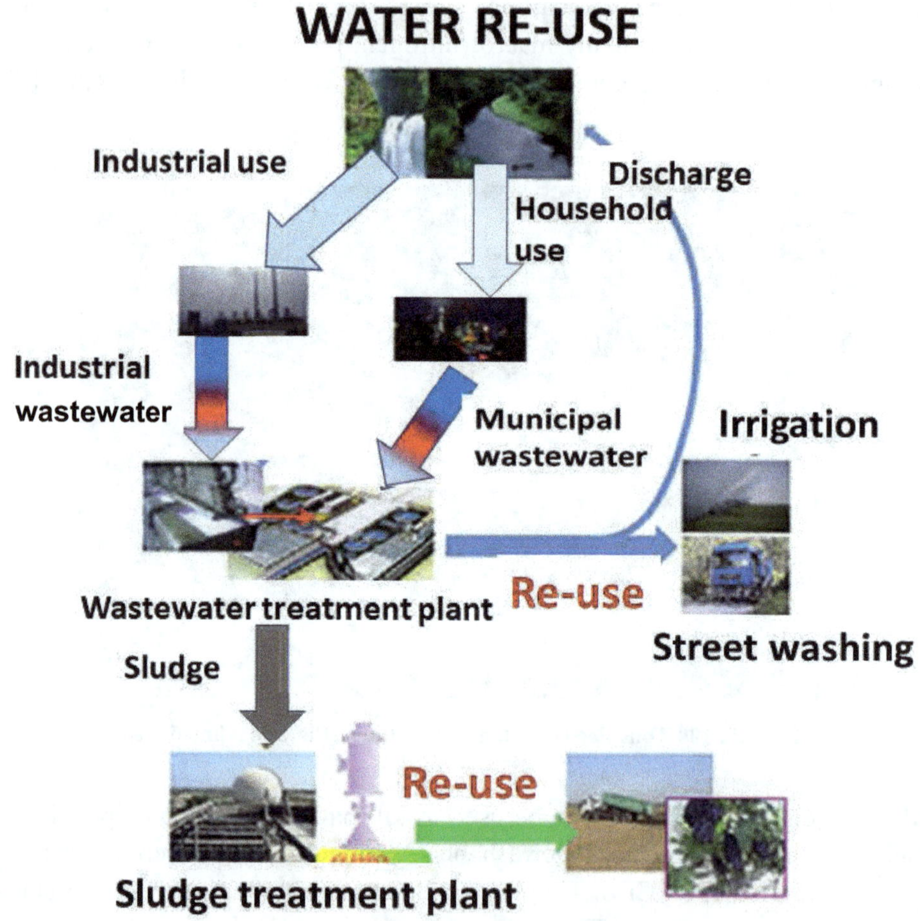

Fig. 15.6: Electron Beam Treatment of Wastewater [206]

Wastewater generated from textile mills, paper mills and tanneries are often discharged directly to natural water bodies and pose great danger to the aquatic flora and fauna. Most of the dyes are not biodegradable and a layer of coloured water does not allow sunlight to penetrate beneath the surface thus inhibiting photosynthesis for plants. It also uses up the valuable dissolved oxygen necessary to sustain aquatic life. Internationally, many experiments have been carried out on EB treatment of wastewater in countries like Russia, Brazil, and South Korea with high power electron accelerators. DC accelerators have been employed for this purpose and a 1 MeV 100 kW electron beam unit can treat up to 2 million litres of water per day, which can then be safely used for irrigation and industrial applications. One such unit, for installation in Mumbai, is under construction.

Radiation is also used to help remove toxic pollutants, such as exhaust gases like sulphur dioxide and nitrogen oxides, from coal-fired power stations and industry. Radiation is used extensively to make wrinkle-free self-cleaning textiles and diapers.

Fig. 15.7: High Yielding Groundnuts Developed by Bhabha Atomic Research Centre [207]

The agricultural industry has been making an extensive use of radiation to improve food production and packaging. The Bhabha Atomic Research Centre has exposed plant seeds to radiation to produce high-yield groundnuts and lentils. In addition, many of our foods are packaged in polyethylene shrink-wrap that has been irradiated so that it can be heated above its usual melting point and wrapped around the foods to provide an airtight protective covering.

Engineers use gauges containing radioactive substances to measure the thickness of paper products, fluid levels in oil and chemical tanks, and the moisture and density of soils and material at construction sites. They also use an x-ray radiography, to locate otherwise imperceptible defects in metallic castings and welds. Radiography is also used to check the flow of oil in sealed engines and the rate and way that various materials wear out. Some devices use a radioactive source and detection equipment to identify and record formations deep within a bore hole (or well) for oil, gas, mineral, groundwater, or geological exploration.

Fig. 15.8: Embankments Made in the Catchment Areas to Revive Springs in Himalayas

Uttarakhand is situated in the lap of the Himalayas and is known as the 'water bank' of the country where some key rivers originate. However, its hill areas have always faced water shortage. With temperatures rising, rivers flowing through deep valleys are unable to meet the needs of parched villages in the hills. Areas in the state that are dependent on natural water resources face a crisis every summer. Destruction of forests has led to drying up of natural water resources that were crucial for the survival of people in the mountains. Bhabha Atomic Research Centre, working with Himalayan Environmental Studies and Conservation Organisation (HESCO) collected water samples from various catchment areas of dried-up springs during the monsoon when there was slight discharge from these springs. The scientists used isotope hydro geo-chemical technique to track the origin of the dried-up spring on the slopes of the hills. Once the origin and the route were traced, they built bunds so that water started percolating. After establishing the catchment area of each spring, water bunds and tanks were set up to hold rainwater in the recharged zones. Once this happened, the spring was recharged at the village downstream. We can also use isotope-hydrology to study contamination of underground water sources.

Radioactive materials also power our dreams of outer space, as they fuel our spacecraft and supply electricity to satellites that are sent on missions to the outermost regions of our solar system, as we have seen earlier.

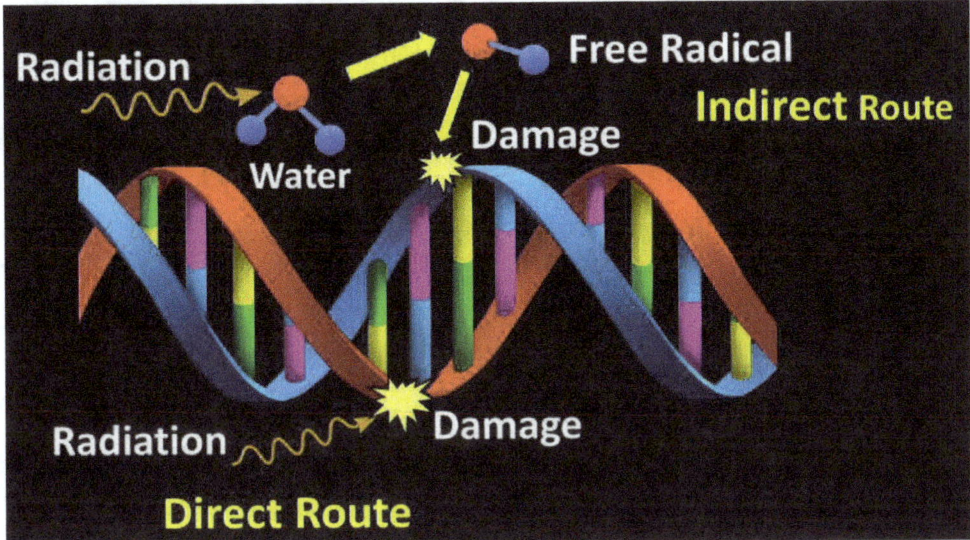

Fig. 15.9: Direct and Indirect Ways to Radiation Damage of DNA [208]

Finally, we add that radiations from x-rays, electron-beams, and Co-60 have been used extensively for the treatment of cancer for decades by now. A high energy radiation or a fast-moving particle traversing a living cell knocks electrons from molecules that make up the cell. The molecules with missing electrons are called ions. The presence of these ions disrupts the functioning of the cancerous cells.

The most severe damage to the cell results when the DNA (deoxyribonucleic acid) is injured. There are two major ways that radiation injures the DNA inside cells.

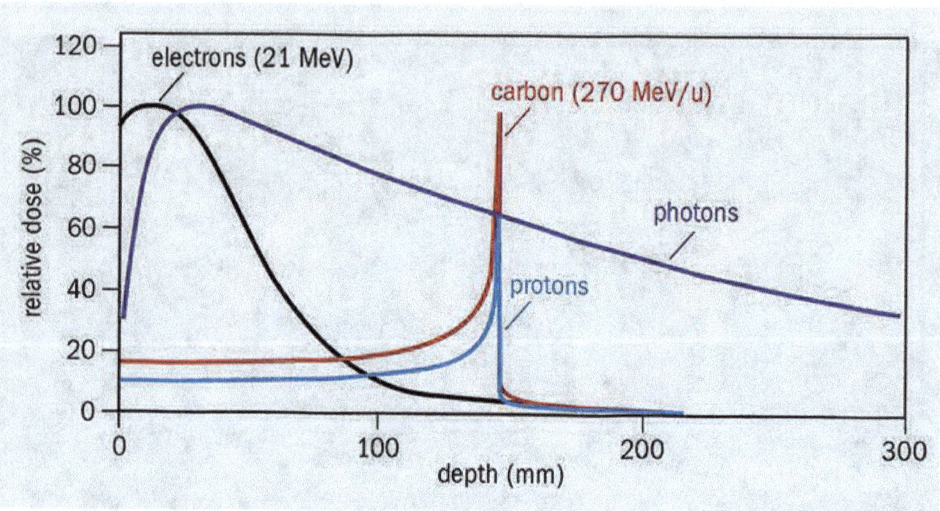

Fig. 15.10: Localized Deposition of Radiation Dose Using Particle Beams [209]

1. The water in the body tends to absorb a large portion of the radiation and becomes ionized. When water is ionized it readily forms highly reactive molecules called free radicals. These free radicals can react with and damage the DNA molecule.
2. Alternatively, radiation can collide with the DNA molecule, itself, ionizing and damaging it directly.

The gamma-radiations from a Co-60 source or x-rays deposit less and less energy as they penetrate the body of the patient. If the cancerous tumour is deep inside the body, the healthy tissues on the way to the tumour are unnecessarily damaged.

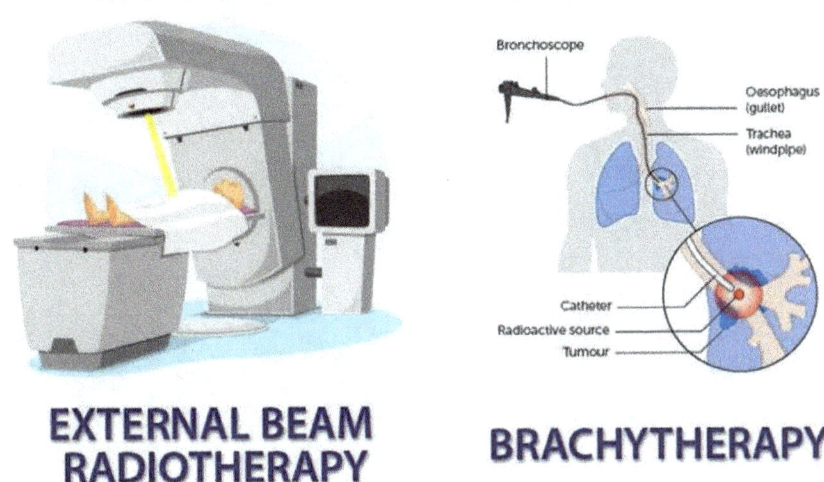

Fig. 15.11: External Beam Radiotherapy and Brachytherapy (IAEA, Facebook) [210]

A very localized dose of radiation without damaging the neighbouring areas, can be delivered by using proton or carbon-12 beams. This procedure is being extensively used, especially if the tumour is close to sensitive organs.

Brachytherapy is another type of radiation therapy used to treat cancer. It places radioactive sources inside the patient to kill cancer cells and shrink tumours. This allows doctors to use a higher total dose of radiation to treat a smaller area in less time. It has been found to be very effective in the treatment of prostate cancer.

Chapter 16

Hydrogen

Before we formally close our discussions on the world energy resources, it is worthwhile to discuss one other source of energy, the hydrogen. We shall also see in the following, that hydrogen, while being a very powerful energy source, is not an energy resource in the same spirit as coal, oil, or natural gas.

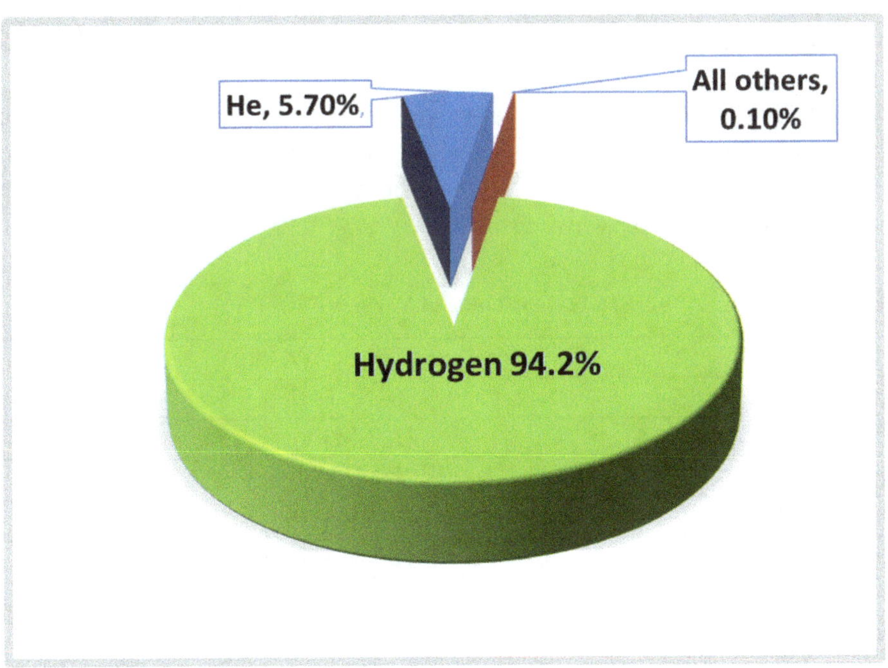

Fig. 16.1: Abundance of Elements in the Universe by Number of Atoms

Produced soon after the Big Bang, the very beginning of the Universe, hydrogen constitutes nearly three quarters of the baryonic mass of the Universe and is the most abundant chemical element in the Universe. The proton–proton fusion reactions leading to the formation of helium release huge amounts of energy in the Sun and other stars and generate all the light and heat that they radiate.

Hydrogen is an essential element for life on Earth. It is present in water and in almost all the molecules in living beings. At present about 70% of the area of the Earth is covered by water, which amounts to about 1.4 billion cubic kilometres by volume. Hydrogen is also present in plant fossils such as coal, petroleum, and natural gas. The amount of pure hydrogen in Earth's atmosphere is indeed very low, typically around 0.6 parts per million. Being very light, hydrogen escapes into the upper atmosphere very fast and eventually escapes into space. Human understanding of hydrogen has come a long way since its production by Robert Boyle in 1671. It was recognized as a distinct element by Henry Cavendish only in 1766. One of the early applications of hydrogen was indeed to lift balloons and airships because of its low density. However, it reacts explosively with oxygen and its practical use in filling airships ended when the Hindenburg airship caught fire (1937). Its explosive nature also makes it an excellent fuel.

From the table (Table 16.1) given below, we can see that hydrogen has the largest heat value of all the fuels per unit mass. It produces water upon burning and thus it is also the cleanest energy source, which holds greatest promise for mitigation of air pollution and global warming.

Table 16.1: Heat Value of Various Fuels [211]

Hydrogen (H_2)	120–142 MJ/kg
Methane (CH_4)	50–55 MJ/kg
Methanol (CH_3OH)	22.7 MJ/kg
Dimethyl ether (CH_3OCH_3)	29 MJ/kg
Petrol/gasoline	44–46 MJ/kg
Diesel	42–46 MJ/kg
Crude Oil	42–47 MJ/kg
LPG	46–51 MJ/kg
Natural Gas	42–55 MJ/kg
Coal	10–25 MJ/kg
Wood (dry)	16 MJ/kg

An early application of hydrogen as a fuel was the Space Shuttle whose main engine was powered by burning liquid hydrogen and liquid oxygen. This choice was basically because of their high energy densities. The nice thing about using hydrogen as fuel is that the exhaust is pure water vapor. It is not surprising that hydrogen is seen as the cleanest fuel.

Fig. 16.2: Space Shuttle Riding a Liquid Hydrogen and Liquid Oxygen Fuelled Rocket [212]

Traditionally, hydrogen is produced by heating natural gas with steam to form syngas (a mixture of hydrogen and carbon monoxide). The syngas is further processed to give pure hydrogen. The process involved adds considerable carbon load to the atmosphere. We should remember, that in this process, one releases 11 kilograms of carbon dioxide for every kilogram of hydrogen, along with other pollutants, e.g., sulphur dioxide, *etc.* We have already seen that coal, biomass, and oil can also be used in the process called "coal gasification" to produce hydrogen. This hydrogen is called "Grey Hydrogen". We shall discuss this in a little more detail, later.

Hydrogen produced using coal, oil, natural gas, biomass, *etc.* by reforming and shifting reactions but enforcing carbon capture and storage — either for industrial use, or underground or in deep sea is referred to as "Blue Hydrogen".

Environmentalists do not recommend 'blue' hydrogen options, as explosive release of the stored carbon dioxide into the atmosphere in future cannot be ruled out. A method of molten metal pyrolysis to separate hydrogen from carbon, is being tried, which produces, what is called "Turquoise Hydrogen".

The hydrogen produced by methods which do not release any carbon into the atmosphere such as electrolysis of water using electricity from renewable energy sources such as Solar PV modules, Windmills, Tidal Power, or Nuclear Power plants, is called "Green Hydrogen".

Hydrogen Storage and Distribution

Hydrogen can be distributed through pipelines or it can be stored as gas under high pressures in specially designed containers or in compressed gas cylinders. It can also be carried in fuel tanks as in the case of transport vehicles like cars and buses. We add that, hydrogen is the smallest element and thus has a slightly higher propensity to leak from venerable natural gas pipes such as those made from iron. However, leakage from plastic (polyethylene PE100) pipes is very low.

Hydrogen can also be stored and transported as liquid at very low temperatures (253 degrees Celsius below zero) using insulated cryogenic tanks and pipes. Japan is in the process of launching a ship to transport liquid hydrogen.

Another interesting possibility is to convert renewable electricity into an energy-rich gas, ammonia, which can easily be cooled and squeezed into liquid, stored, and transported anywhere. The ammonia can be converted back into hydrogen by fracking and to produce electricity at the point of use.

Adsorption of hydrogen in some metals like palladium or conversion into hydrides, for transporting hydrogen (as a solid) is also being tried.

Uses of Hydrogen

In the chemical industry, hydrogen is used to make ammonia for agricultural fertilizers and cyclohexane and methanol which are intermediates in the production of plastics and pharmaceuticals. It is also used to remove sulphur from fuels during the oil refining process. Large quantities of hydrogen are used to hydrogenate oils to fats, for example, to make margarine. There are also other industrial applications.

Hydrogen can be used in much the same way as natural gas — to generate heat which can be used in a combined cycle gas turbine to produce larger quantities of centrally produced electricity. This will produce no pollutants and no green-house gases, as the product is water. This holds out the promise of being a valuable and effective tool in the mitigation of the crisis of climate change.

Hydrogen can directly drive internal combustion engines or can produce electricity in a fuel cell which can be used to drive an electric engine. The automobile sector is recognized as one of the major polluters of the atmosphere. The world has more than one billion cars on the road and its oil resources are not likely to last more than 50 years. The transport sector is responsible for the emission of about 15% of the greenhouse gases. Electric cars are emerging as a viable alternative to the automobiles powered by hydrocarbons like petrol and diesel. While these help in the reduction of atmospheric pollution, they do not help in the crisis of climate change if the electricity for charging their batteries is generated using fossil fuels.

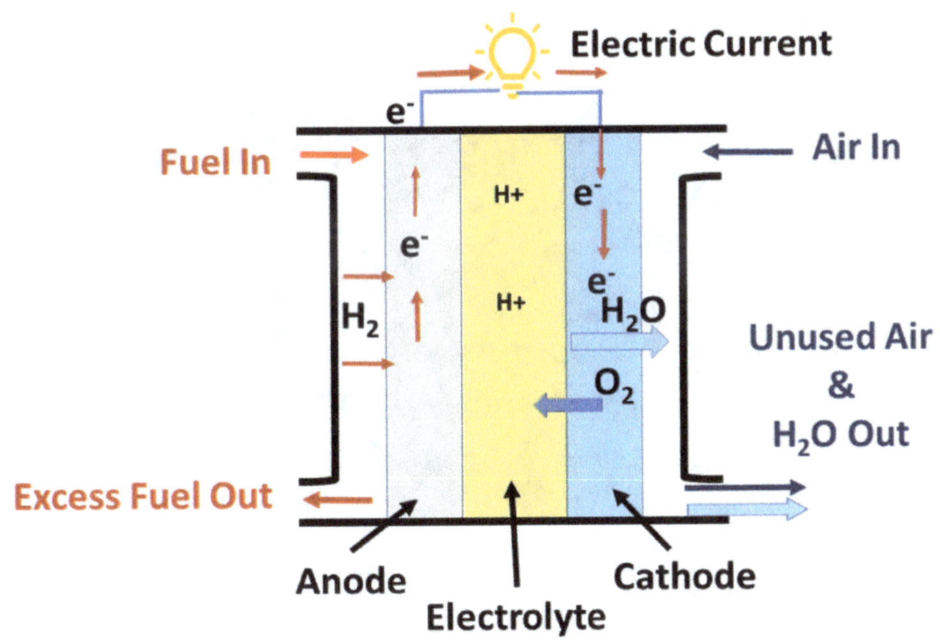

Fig. 16.3: Schematic Diagram of a Typical Fuel Cell [213]

A major drawback of the present generation electric cars is their use of batteries for storing electricity. The batteries not only have a limited range and life but also involve in their manufacture several rare materials — sourcing which may involve their own environmental damage. One exciting possibility, as indicated above, is the use of hydrogen to run cars. Hydrogen can be used directly in internal combustion engines with necessary modifications. An alternate technology is to use fuel cells to generate electricity and use the electricity to drive an electric car.

Fuel cell uses hydrogen and oxygen to produce electricity, heat, and water and is thus an electrochemical energy conversion device. It consists of an anode, cathode, and an electrolyte membrane. In a typical fuel cell hydrogen is passed through the anode and oxygen is passed through the cathode. The hydrogen molecule is split into electrons and protons at the anode. The electrons pass through a circuit, leading to the production of heat and electricity. The protons cross through the porous electrolyte membrane and reach the cathode, where they combine with electrons and oxygen to produce water molecules.

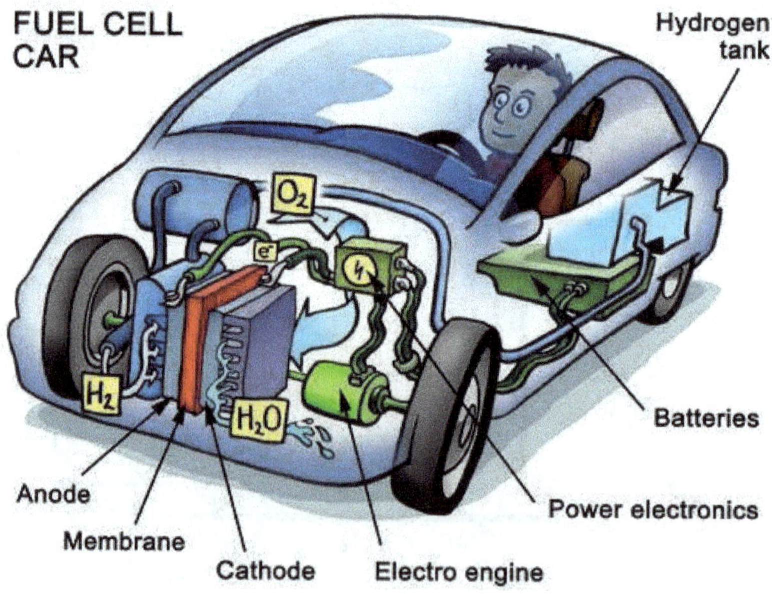

Fig. 16.4: A Hydrogen Fuel Cell Car]214]

Fuel cells do not have any moving parts and operate silently and with extremely high reliability. These do not need to be recharged from time to time, like batteries and continue to produce electricity if the fuel is available. This makes them a clean, efficient, reliable, and quiet source of power.

Hydrogen Road Map

The Ministry of New and Renewable Energy, Government of India has set up a National Hydrogen Energy Board, to accelerate the development and utilization of hydrogen energy in the country. A National Hydrogen Energy Road Map was prepared with the help of high-level representation from Government, Industry, Research Institutes, and Academia, and adopted by the Institutions and Academia and the Government in 2006.

The road map laid out the plans for research and development for:

- production of hydrogen,
- development of internal combustion engines, fuel cells, and
- storage and transport of uncompressed and compressed as well as liquid hydrogen for its effective utilization.

We note that several countries in Europe plan to use hydrogen obtained from hydrolysis of water using the excess electricity produced by windmills to run trains, trams, and buses and the prototypes for these are already available. These will, thus be the greenest means of transport.

Small ships and planes running on hydrogen are already operating. In an interesting initiative, the Government of Delhi has promised to run a fleet of buses where up to 10–30% of the CNG being used as fuel is replaced by hydrogen by December 2020. Cars, though not yet many, are also plying to either use hydrogen in an internal combustion engine or in their fuel cells. Their large-scale entry will require setting up an extensive network for the distribution of hydrogen.

Fig. 16.5: The National Hydrogen Energy Road Map, Government of India, 2006 [215]

Fig. 16.6: Protype of a Train Running on Hydrogen [216]

Several other sectors like steelmaking, heating, and chemicals are beginning to develop large-scale hydrogen applications to gradually replace fossil fuels.

Fig. 16.7: Hy-4: First Passenger Aircraft Powered by a Hydrogen Fuel Cell [217]

The world consumes about 70 million metric tons of pure hydrogen and about 30–40 metric tons of hydrogen mixed with other gases, every year. Natural gas accounts for 48% of this production, oil accounts for 30%, coal for 18%, and electrolysis for 4%.

Hydrogen, as we saw, is often called green, grey, blue, or turquoise depending on its method of production.

Green hydrogen uses renewable energy such as Hydroelectricity, Windmills, Solar Photovoltaics, Tidal/Wave Power, or Nuclear Power for electrolysis of water for its production. There are several ongoing efforts to use offshore windmill farms and deploy floating electrolysis plants to facilitate this. Electrolysis, as the name suggests, is the process of using electricity to split water into hydrogen and oxygen.

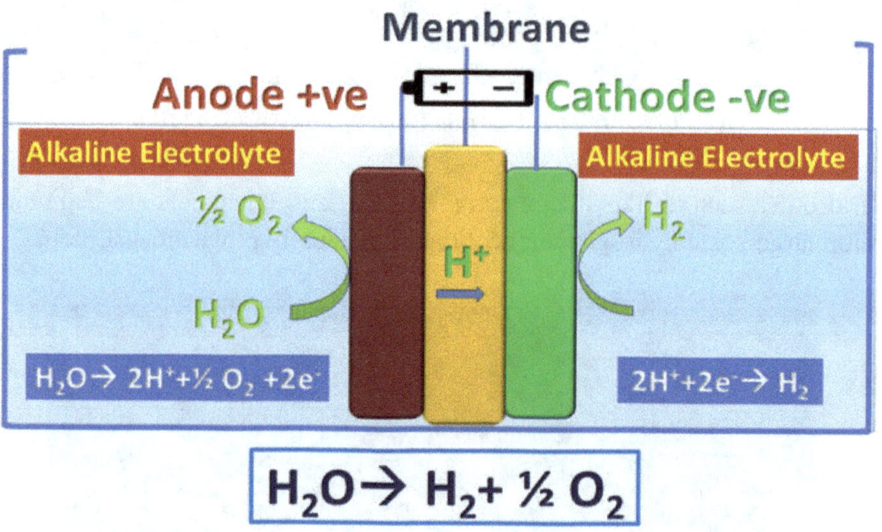

Fig. 16.8: Basic Scheme of Hydrolysis of Water using a PEM Electrolyzer [218]

Electrolyzers consist of an anode and a cathode separated by an electrolyte. Several electrolyzers have been developed which use different electrolytes.

Polymer Electrolyte Membrane (PEM) electrolyzer, for example, uses a special plastic material and works as follows:

- Water reacts at the anode to form oxygen and positively charged hydrogen ions (protons).
- The electrons flow through an external circuit and the hydrogen ions selectively move across the PEM to the cathode.
- At the cathode, hydrogen ions combine with electrons from the external circuit to form hydrogen gas. Anode Reaction: $2H_2O \rightarrow O_2 + 4H^+ + 4e^-$ and Cathode Reaction: $4H^+ + 4e^- \rightarrow 2H_2$.

By now, large capacity electrolyzers having efficiency of about 80% have been developed. The hydrogen and oxygen obtained from electrolysis are very pure (> 99.9%), which is also ideal for some high value-added processes such as the manufacture of electronic components, e.g., silicon chips. If the efficiency of the electrolyzer is about 70–75%, it will need about 50–55 kwh of electricity to produce one kilogram of hydrogen.

Nuclear power plants, especially the ones operating at very high temperatures, produce high-quality steam. This high-quality steam can be electrolyzed to produce pure hydrogen and oxygen. It is estimated that a single 1,000-Megawatt nuclear reactor can produce 200,000 tons of hydrogen every year. High-quality steam at high temperatures can also be produced at Solar Concentrator Plants for this purpose.

Hydrogen extracted from fossil fuels such as coal-using coal gasification or gas-using steam-methane reforming and releasing carbon monoxide and carbon dioxide is called *grey hydrogen*. We have seen that in coal gasification, we first have a reaction between coal, oxygen, and steam

$3C\ (\textit{as Coal}) + O_2 + H_2O \rightarrow H_2 + 3CO$

leading to 'syn gas' and then we have the shift reaction between carbon monoxide and steam

$CO + H_2O \rightarrow CO_2 + H_2$.

While using methane, we first have the reform reaction between methane and steam

$CH_4 + H_2O$ (+ heat) $\rightarrow CO + 3H_2$.

This is followed by the shift reaction as before. Production of one kilogram of grey hydrogen is accompanied with the release of eleven kilograms of carbon dioxide. The use of coal will also lead to release of sulphur dioxide and other pollutants, including particulate matter.

If this carbon dioxide is separated (captured) for reuse or for underground storage or subsea storage, then this hydrogen is called *blue hydrogen*. This is seen as a transitional approach, till the demand for the hydrogen cannot be met by green hydrogen. However, as mentioned earlier, this is opposed by environmentalists as this method of storing carbon dioxide is fraught with the danger of its explosive release.

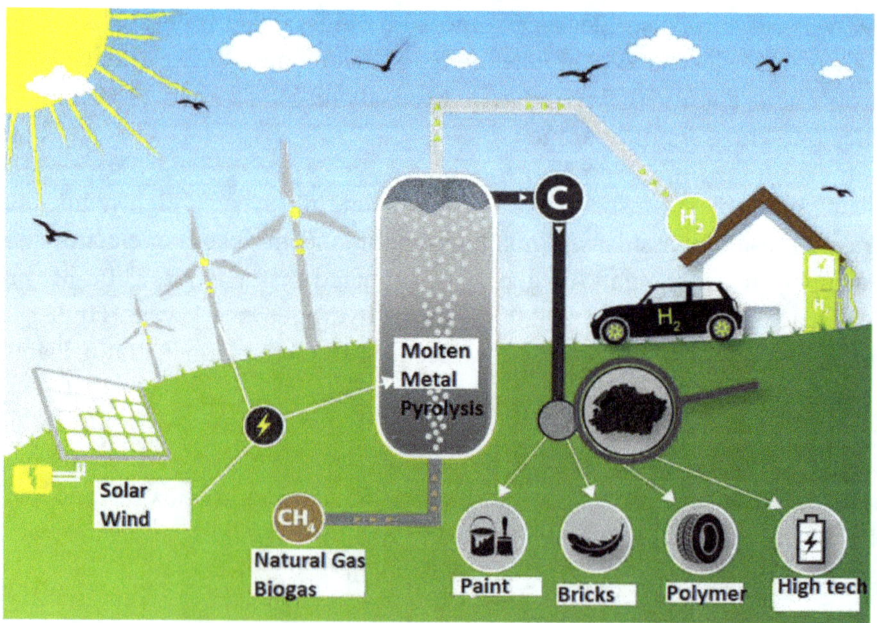

Fig. 16.9: Turquoise Hydrogen [219]

An interesting concept of low-carbon hydrogen, called *turquoise hydrogen* involves pyrolysis of natural gas by passing it through molten metal, which produces solid carbon as a useful by-product and hydrogen. So far this is available only at a small scale, but plans are afoot to deploy it on a large scale.

As indicated earlier, some microbes, such as green algae, consume water in the presence of sunlight and produce hydrogen. An interesting new concept involves using photoelectrochemical splitting of water to produce hydrogen using special semiconductors and energy from sunlight.

Hydrogen Hamlets

India is a country of countless villages. More than 75% of the population lives in non-metropolitan settlements. Having a national hydrogen network is indeed a Herculean task, not only from the point of view of resources but also from the point of view of effort. It has been suggested that renewable sources of energy, like solar photovoltaics, windmills, running water, *etc.* coupled with hydrogen storage and fuel cells can be a very reliable source of electricity for isolated hamlets and can also provide transport fuels.

Fig. 16.10: Hydrogen Economy for Isolated Hamlets [220]

We conclude that, even though still in its infancy, hydrogen holds out the promise of storage and supply of very clean energy in years to come. In fact, it has been suggested that the most abundant gas in the Universe could help the most abundant energy poor on Earth, lift themselves out of their abject poverty.

Chapter 17

Summary and Outlook

Summary

We have seen that the Earth is the only planet in the Solar System and perhaps the only one that we know, which supports life in all its wondrous variety and beauty. The human species is a very recent addition. Starting from a scattered population of small groups in Africa, they now occupy the entire world today and their number exceeds 7.5 billion. Along the way, they have made discoveries in sciences and mastered more and more technologies to make their lives more comfortable and fulfilling. This also led to the emergence of several vibrant civilizations, with their ups and downs, and yet marching forward all the time.

Apart from food and water, this progress also needed energy, way beyond what their muscles could provide. Starting from domesticated animals, running water, and blowing wind, humans have benefitted substantially from the use of biomass, coal, oil, and natural gas. The discovery of electricity has substantially widened the scope of energy resources such as hydroelectricity, solar energy, wind energy, tidal energy, geothermal energy, and nuclear energy. Hydrogen gas obtained from electrolysis of water using renewable energy is fast emerging as a carrier of green energy to be used for production of electricity, use in industry, transportation, and to supplement or replace natural gas.

We have already seen that the economic strength of a nation as reflected in their GDP and the quality of life of its citizens as reflected in their Human Development Index depend rather sensitively on the per capita availability of energy. It is estimated that an electricity consumption of about 4000 kWh/year per person should provide conditions for a comfortable and fulfilling life for the people. On the other hand, the fraction of the world population having access to energy of this order and having a comfortable life is indeed small. A substantial fraction of the population has little access not only to electricity but also to education, primary health care, and other essential services. About 1 billion people still do not have access to electricity at all. And yet, for the first time in the history of mankind, science and technology have provided us with knowledge and means to remove all these sufferings. To do this, we need to very rapidly jack-up energy availability.

In the last two centuries, the world has depended largely on fossil fuels like coal, oil, and natural gas to meet up to 80% or more of the global energy needs. This has led to unprecedented levels of carbon dioxide (more than 410 ppm in 2019) in the atmosphere. That, in turn, has led to a rise in global temperature, climate change, rise in sea level, rise in sea temperature, acidification of oceans, melting of ice in Greenland and the poles — along with the danger of associated release of methane from the melting of permafrost, death of glaciers, drying up of snow-fed rivers, repeated and prolonged droughts, forest fires, flash floods, disruption of the pattern of monsoons, destruction of the habitat of fish, increased cyclonic activities, and crop loss, *etc.* The requirement of increasing food has led to large scale deforestation, which further exacerbates the situation.

It has been abundantly clear for many decades that the increasing demand for energy must not be met by fossil fuels, which in any case may not last for more than 50–100 years to come. It is feared that if we continue to burn fossil fuels at the rate that we are burning, the continuation of our very existence could be very much in doubt.

This realization has led to increasing deployment of green energy sources, though the pace is not at all enough to limit the rise in temperature to less than 1.5 to 2.0 degree Celsius recommended by Intergovernmental Panel on Climate Change. This requires reduction of Green House Gas emissions to 40% of that in 2010, by 2030 and to zero by 2050.

It is also widely realized that the only thing that will save us and future generations from paying a huge price in terms of money, lives, and damage to nature is rapid and substantial reductions in carbon emissions from fossil fuels, to net zero by 2050.

It is in this context that the emergence of green sources of energy like wind energy, solar photovoltaics, solar thermal, and nuclear energy assume significance.

Outlook

We have seen that the entire world is struggling today with the problem of providing enough reliable and affordable power to all. The challenge is to do this without exacerbating the already precarious state of global climate, which is threatening our very existence. A very optimistic vision of the entire energy needs of select 139 countries being met by renewables like solar, wind and other sources by 2050 has been put forward by some researchers.

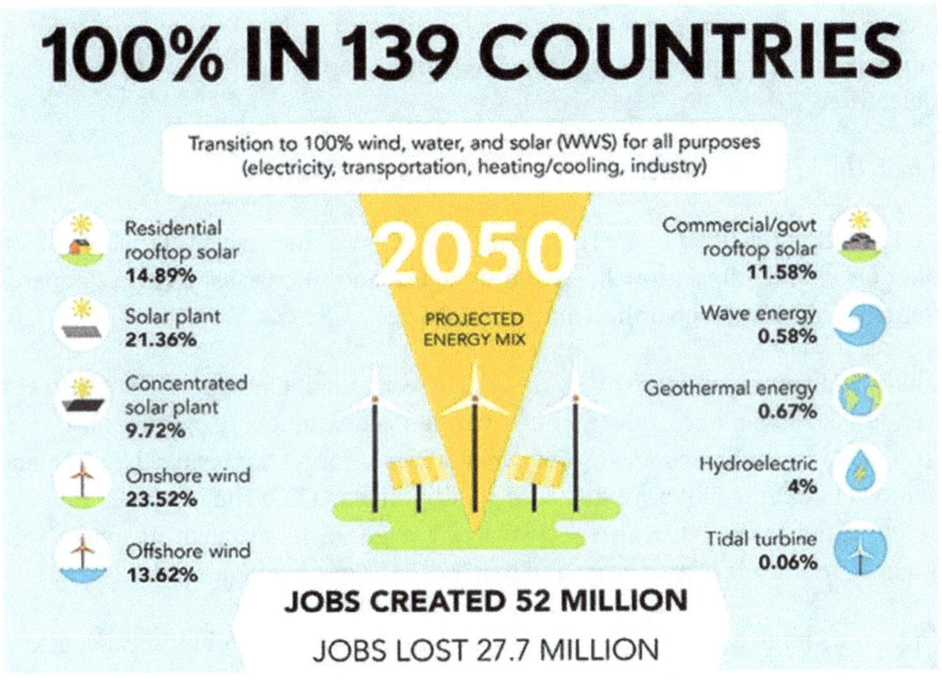

Fig. 17.1: A Vision for 100% Renewable Energy for 139 Countries [221]

Can the distributed power, that too with day-night and strong seasonal variations meet the demand of heavy industries like production of steel or cement as we know today? Storage of electricity using batteries with the required capacities is still a long way off, even though the possibility of using the renewable energy to produce hydrogen by electrolysis as an energy storage is looking very promising and it is being implemented widely, as mentioned above.

Possibility of a global renewable electricity grid to address the day-night variations of solar electricity has already been in the air for close to a century. R. Buckminster Fuller (American thinker, 1895–1983; see [222]) developed the World Game simulation in 1960s, posing the question:

"How do we make the world work for 100% of humanity in the shortest possible time through spontaneous cooperation without ecological damage or disadvantage to anyone?"

Research shows that the premier global strategy is the interconnection of electric power networks between regions and continents into a global energy grid, with an emphasis on tapping abundant renewable energy resources — a world wide web of electricity.

In fact, Buckminster Fuller [223] went on to suggest that:

"Linking the renewable energy resources around the world reduces global pollution, population growth, world hunger, and increases living standards, international trade, cooperation and world peace."

India has taken an initiative by starting a project called, One Sun One World One Grid, to take advantage of the fact that the Sun is continuously bathing half of the Earth with its light and heat. Yet we must not forget that densely populated countries like India will have difficulties in finding suitable sites for the renewable energy stations. We have also seen that the material requirement to make solar panels or windmills is also large and may pose serious challenges.

When Buckminster Fuller first discussed the concept of global energy grid in 1930s, the nuclear power was nowhere on the horizon. By now we know that, despite the safety concerns often expressed against nuclear electricity, it remains one of the most reliable and cleanest sources of plentiful electricity. We have already seen that, even with the present level of development and the available uranium and thorium resources globally, it can provide reliable power to the entire world for thousands of years.

It is also important to recognize that these days nuclear power plants operate under very stringent safety standards, which are continually updated to consider the latest research on nuclear safety, under the watch of the International Atomic Energy Agency.

Fig. 17.2: Sheikh Saadi (1210–1292) in Rose Garden [224]

A global renewable electricity grid can be very robust and rapidly grow to meet the challenge of growing energy demand of the world if it can be coupled to nuclear power stations, without adding any carbon burden. This strategy also offers the possibility of locating the nuclear power stations at sites chosen based on safety rather than based on user proximities. Such a global superpower grid will also enable safe and secure siting of nuclear power stations. Precedents exist in the international transportation and telecommunication sectors.

The entire strategy outlined above calls for international collaborations at levels far beyond today's ground reality. At the same time let us not forget that the world has always come together when there is a need.

Sheikh Saadi (1210–1292) of Persia, writing eight centuries ago, declared in his famous poem, Bani Adam (Chapter 1, Story 10, Gulistan):

"The sons of Adam are limbs of each other,

Having been created of one essence.

When the calamity of time affects one limb,
The other limb cannot remain at rest.
If thou hast no sympathy for the trouble of others,
Thou art unworthy to be called by the name of a human."

(Motto engraved at the entrance of United Nations Building)

Let there be no doubt that the only way to stop our beautiful planet from overheating and seriously affecting or even completely jeopardizing the survival of human civilization is through political, economic, technological and social solutions that end the use of fossil fuels.

Let us also be very clear that carbon-free power must be adopted by every nation in the world. The burdening of the atmosphere by carbon dioxide may have been predominantly done by only some of the industrialized nations, but the climate change and global warming affects every nation in the world. This coupled with the urgent need to pull the entire mankind from the grip of poverty and suffering will require efforts by every nation. No nation is isolated, when it comes to global warming or even pollutions of oceans and other waterbodies and the air we breathe.

It will need global, national, and local action. For this we shall need not only stable and coherent government policies but also, perhaps even more importantly, well informed, and discerning public.

We recall that the composers of Mahaupnishad (6.71-75) declared:

"अयं निजः परो वेति गणना लघुचेतसाम्।
उदारचरितानां तु वसुधैव कुटुम्बकम्॥"

"This is mine, that is his, say the small minded. To the wise, the entire world is a family."

The choice is ours to make.

Chapter 18

Epilogue

Let us start with a story with a powerful message. Mahatma Gandhi went to England in 1931 along with leaders of various political parties and several Rajas and Maharajas, to participate in the Round Table Conference and negotiate India's independence from Britain. He was invited to meet King-Emperor George-V.

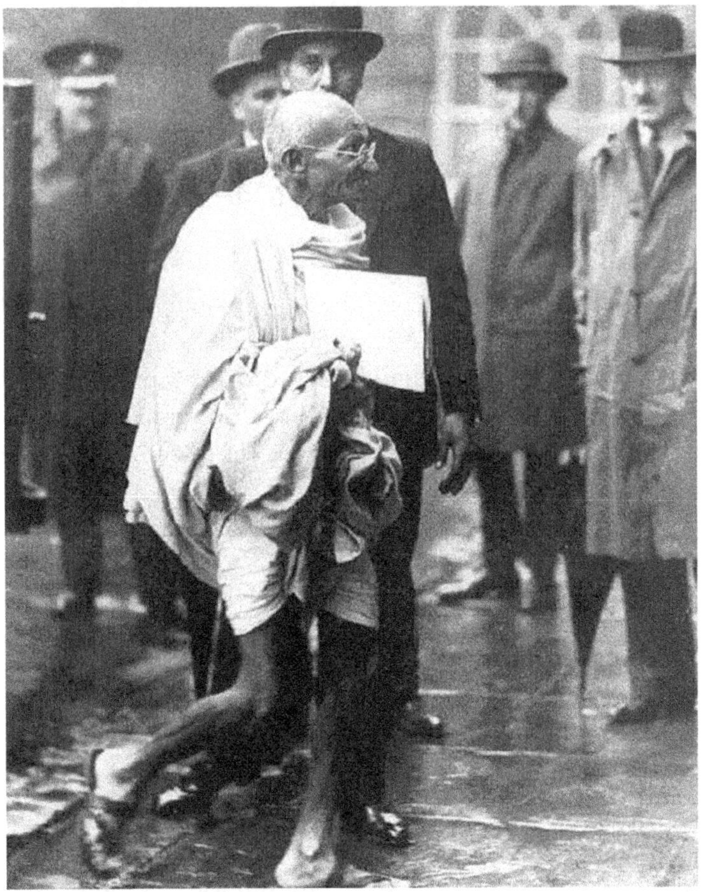

Fig. 18.1: Mahatma Gandhi Arrives to Meet King-Emperor George-V (1931)

Mahatma Gandhi arrived at the Buckingham Palace, in his loincloth, sandals, bare legs, and a rough woollen shawl wrapped around his upper body. The palace officials were aghast at his attire. When the meeting was over and he was walking out of the palace gates, a journalist asked him if he thought he was wearing enough. Mahatma Gandhi with his usual simplicity and humour is said to have replied, "But the King was wearing enough for the both of us". Given the predicament that we find ourselves in today, we shall have occasions to return to Mahatma Gandhi.

It has been estimated by the international research consortium of scientists, "Global Carbon Project", that global emission of carbon dioxide from burning fossil fuels for 2019 was 37 billion tons. This corresponds to burning 14,800 billion litres of petrol. It would take approximately 1,480 billion additional adult trees on our planet to absorb this carbon dioxide in one year, as each tree absorbs about 25 kilograms of carbon dioxide in a year. On the other hand, we have been cutting down about 3.5 to 7 billion trees every year to plant crops or oil palms or for creation of grazing pastures for the meat industry, in addition to those being felled for the traditional timber industry. It is even widely advertised that grass-fed cattle (beef to be specific) provide better meat, even though there is no scientific justification or evidence for this. This has led to the elimination of vast tracks of forests in Argentina, for example.

The systematic devastation of Amazon forests — lovingly named "the lungs of the Earth" is only too well known, and it is getting accelerated. We have also seen the devastating forest fires in California and Australia, burning down hundreds of thousands of acres of forests and killing billions of animals. And adding to our difficulties is the fact that an additional 6 billion tons of carbon dioxide are emitted by agriculture and land use.

This has been the trend for the last several years, to the consternation and despair of scientists studying climate. Every additional ton of carbon dioxide in the atmosphere of the Earth is making our avowed goal of keeping the global warming to less than 1.5–2 degree Celsius and the bringing down of the emission of net carbon dioxide to zero by 2050 by switching to renewables and nuclear power, that much more difficult.

Let us pause here and look back at the development of this problem. We realize the need to understand the impact of our energy consumption on our planet that has limited resources and the impact of the by-products of our energy use on the ecological balance of the Earth. By now it is generally believed that the health of human civilization and the state of the natural systems on which it depends is at serious risk. This risk has its origin in the failure of climate change mitigation and adaptation; extreme weather events; major biodiversity loss and ecosystem collapse; food crises; and water crises.

Several Gandhian thinkers have written extensively on this. Recall Mahatma Gandhi's remark, "The Earth has enough for everybody's need but not for one man's greed". It has been argued that if Mahatma Gandhi were alive today, it is very likely that he would have attributed the problems of pollution, acid rain, increase in carbon dioxide levels in the atmosphere, global climate change, fish dying, and Amazon forest burning, to just one root cause — greed.

It is well known that villages across the world prior to the industrial revolution were essentially self-reliant communities, surviving harmoniously in synchronization with local resources. A typical Indian village was no exception with a way of life developed over thousands of years living harmoniously in synchronization with local resources. Greed was indeed frowned upon.

Industrial revolution changed all that. Western societies could produce more, and worse, produce even when it was not needed. This required the creation of a society that wanted more and more material possessions and pleasure and a culture of hoarding. Driven by this notion of wealth creation, West turned into a furnace that started burning the planet faster and faster and even saw it as a great achievement that it needed to push forward.

This made them look for more and more resources which led to the colonization of other nations. This has had disastrous consequences for the latter — as post-industrial revolution Occidental civilizations perceived life as a constant quest for gratification by an individual, while Oriental civilizations, as we indicated earlier, saw life as a coexistence in sync with fellow human beings and fellow life forms.

Large scale industrialization in many of the western parts of the world not only led to enormous manufacturing capabilities but also created a chronic demand for resources and a consumer market. Colonization for resources and a market for goods manufactured were two of the saddest outcomes of this development.

The World Watch Institute, Washington D. C., has observed that about 1.7 billion people worldwide now belong to the "consumer class" — the group of people characterized by diets of highly processed food, desire for bigger houses, more and bigger cars, higher levels of debt, and lifestyles devoted to the accumulation of non-essential goods.

Nearly half of global consumers reside in developing countries, like China and India. These also represent markets with the largest potential for expansion. The World Watch Institute further reports that the rising consumption has helped create new jobs and meet basic needs.

This unprecedented consumer appetite is however undermining the natural systems we all depend on. The consumerism has taken a devastating toll on the Earth's water supplies, natural resources, and ecosystems. These are caused by, but not limited to, mobile phones, plastic garbage bags, and other cheaply made goods with built-in product-obsolescence, and cheaply made manufactured goods that lead to a "throw-away" mentality.

Egged on by constant beaming of images of consumer goods, vacations at exotic places, foods from across the world — made to travel thousands of kilometres to reach their tables, the world is consuming resources like energy and water, at a rate never before witnessed in the history of mankind.

We have already noted that western nations use up to 12,000 kWh/year per person, while just one-third of it can provide Human Development Index of about 0.8. The Index reaches a plateau around this level of energy consumption.

To emphasize the culture of wastage, we note that it is known that on an average a person in the West, especially the USA, throws away 40 kilograms of clothes per year. A large part of it is due to the emergence of "Fast Fashion" as a marketing strategy. Often it is simply dumped into landfills, where it may take hundreds of years to decompose.

Also recall that for Fast Fashion to succeed, very cheap labour in countries like Bangladesh or Vietnam is needed. To stay competitive, the manufacturers are forced to adopt unsafe working conditions and release pollutants in the environment. The April 24, 2013 collapse of Rana Plaza building in Dhaka, Bangladesh killed 1,132 people and injured more than 2500. The building housed five garment factories. There are reports that the suppliers had to bargain cent by cent per piece, over things like buttons and zips, to secure orders.

Visions of being fashionable, liberated, and independent — aided by a sustained advertising using glamorous models, were used to spur women to start smoking and double the sales of cigarettes in the USA and the West. Introduction of a new model of car or mobile phone is used to create artificial demand.

All these have led to a lifestyle of excessive materialism that revolves around reflexive, wasteful, or conspicuous overconsumption. We add that "conspicuous consumption" denotes a situation where consumers purchase, own, and use products not for their direct use but as a way of signalling their exalted social and economic status.

It has been suggested that several manufacturers also resort to planned or contrived obsolescence to limit the durability of a product, to increase sales.

This could involve a deliberate use of inferior materials in critical areas, which cause excessive wear. The short life expectancy of smartphones and other handheld electronics is believed to be a result of constant usage, fragile batteries, and the ability to easily damage them. It is making it even harder for the world's poor to meet their basic needs.

The conspicuous consumption uses precious resources at a rate which cannot be replenished by the Earth. When it comes to food, at least, the Earth with the help of Sun, struggles to provide enough to us every year. There are however estimates that up to 40% of the food, vegetables, and fruits produced are wasted every year for want of storage facilities.

Occasionally, bumper productions are deliberately destroyed to stabilize the prices. At the other end of the spectrum, the world still has a large population of starving, malnourished, growth stunted people, not because there is not enough food going around but because they cannot afford it.

We are rapidly running out of mineral resources and are looking at new fields such as the Antarctica or deep oceans for mineral deposits. Knowing the track record of mining industries across the world over centuries, this is quite likely to leave deep scars on the face of the Earth.

Disposal of industrial products at the end of their life is yet another challenge that we need to confront. Almost all major cities have scrap yards for metals, which provide one of the saddest sights not only from the point of view of loss precious mineral resources but also from the point of view of human labour which was used in creating the discarded objects.

We end this discussion by re-emphasizing that conspicuous consumption by humanity has done severe harm to the delicate balance of nature and environment and it is reaching a tipping point where our own existence is at stake. It is also important that the vast population of the Earth realizes the disastrous path which they have been following. Fortunately, Science and Technology still point a way out of this morass, if deployed across the globe cutting across national boundaries.

These are but just a few examples, which we have discussed to show the unnecessary waste of our resources, spurred by greed and a false sense of gratification of senses — which are the most potent tools in the field of advertising and sales. There is enormous literature available on this topic, which invariably overlooks the damage to the environment, due to over-exploitation and wastage.

Gita (Chapter 2, verse 62 and 63) famously declares:

"ध्यायतो विषयान्पुंसः सङ्गस्तेषूपजायते।
सङ्गात् संजायते कामः कामात्क्रोधोऽभिजायते।।2.62।।
क्रोधाद्भवति संमोहः संमोहात्स्मृतिविभ्रमः।
स्मृतिभ्रंशाद् बुद्धिनाशो बुद्धिनाशात्प्रणश्यति।।2.63।।"

"Contemplation on sense-objects leads to attachment. From attachment arises desire. From such desire arises anger. From anger, there comes delusion. Then there follows the failure of memory.

From the loss of memory there comes the destruction of discrimination. From the destruction of discrimination, one becomes lost, i.e., is sunk., in worldliness."

We cannot describe the role of greed in the predicament that the world finds itself in, in a better manner. We just recall that the philosophy of not accumulating things other than what we truly need called aparigraha (अपरिग्रह), forms a very important guiding principle in Vedic, Jain, and Thirukkural traditions of India.

Bibliography

[1] http://www.sahajayogaportal.org/en/music-arts/indian-classical-music/seven-notes.html

[2] Jessie Eastland, Wikimedia Commons

[3] https://commons.wikimedia.org/wiki/File:Eucalyptus_deglupta-trees.jpg

[4] http://avax.news/pictures/110539; Photo by Alberto Ghizzi Panizza/National Geographic Photo Contest 2014, week 8

[5] https://commons.wikimedia.org/wiki/File:Adult_male_Royal_Bengal_tiger.jpg

[6] https://commons.wikimedia.org/wiki/File:Indian_Gaur_from_anaimalai_hills_JEG5290.jpg

[7] https://upload.wikimedia.org/wikipedia/commons/c/ce/Raggiana_Bird-of-Paradise_wild_5.jpg

[8] https://commons.wikimedia.org/wiki/File:Papilio_ulysses_ambiguus_Rothschild,_1895.JPG

[9] https://www.nationalgeographic.com/news/2016/09/humpback-whale-endangered-list-pictures/, Photograph by Ralph Lee Hopkins, National Geographic Image Collection

[10] https://commons.wikimedia.org/wiki/File:Dahlia_x_hybrida.jpg

[11] Jeff Dahl, Wikipedia

[12] https://www.thesacredscience.com

[13] Illustration: Sarada Natarajan; See also, https://rodadoarcoiris.wordpress.com/2018/02/22/criancas-da-nova-era-cristal-e-diamante/

[14] Parthenium hysterophorus, Ethel Aardwark, Wikipedia

[15] Prosopis juliflora, Thamizhpparithi Maari, Wikipedia

[16] Wouter Hagens, Wikipedia

[17] Tannin, Wikipedia

[18] Daiju Azuma, Wikipedia

[19] Richard Taylor, Wikipedia

[20] National Museum, New Delhi

[21] https://commons.wikimedia.org/wiki/File:Maharajah_Ramanuj_Pratap_Singh_Deo_with_cheetah_kill_1948_BNHS.jpg

[22] https://commons.wikimedia.org/wiki/File:Passenger_pigeon_shoot.jpg

[23] https://commons.wikimedia.org/wiki/File:Apis_mellifera_-_Prunus_padus_-_Keila.jpg

[24] https://chinadialogue.net/en/food/5193-decline-of-bees-forces-china-s-apple-farmers-to-pollinate-by-hand/

[25] Illustration: Sarada Natarajan, See also; https://www.sciencelearn.org.nz/resources/890-what-is-in-soil

[26] https://www.unenvironment.org/interactive/beat-plastic-pollution/

[27] Courtesy: NASA

[28] See, e.g.; https://www.visionlearning.com/en/library/Chemistry/1/Early-Ideas-about-Matter/49/reading

[29] https://commons.wikimedia.org/wiki/File:Periodic_Table_Chart.png

[30] St. Petersburg, Russian Wikipedia

[31] See, e.g.; https://www.britannica.com/science/Rutherford-model

[32] https://www.fnal.gov/pub/today/images/images11/NS_Figure01_2011_11_18.jpg

[33] See e.g., https://www.pbs.org/deepspace/timeline/

[34] http://oceanexplorer.noaa.gov/explorations/04fire/logs/hirez/champagne_vent_hirez.jpg, https://commons.wikimedia.org/wiki/File:Champagne_vent_white_smokers.jpg

[35] See, e.g.; https://energyeducation.ca/encyclopedia/Cross_section_of_the_Earth, https://www.researchgate.net/figure/Cross-section-of-the-Earth-showing-the-core-mantle-and-crust-Diamonds-are-generated_fig1_319010186

[36] A loose necktie, Wikipedia

[37] Copyright: The University of Waikato Te Whare Wānanga o Waikato. All Rights Reserved. www.sciencelearn.org.nz

[38] Courtesy: NASA

[39] https://simple.wikipedia.org/wiki/Continental_drift, http://www.geologyin.com/2018/02/facts-about-pangaea-most-recent.htmlhttp://www.geologyin.com/2018/02/facts-about-pangaea-most-recent.html

[40] https://commons.wikimedia.org/wiki/File:Glossopteris_sp.,_seed_ferns,_Permian_-_Triassic_-_Houston_Museum_of_Natural_Science_-_DSC01765.JPG, https://science.howstuffworks.com/life/botany/10-plants-lost-to-history5.htm

[41] https://pubs.usgs.gov/gip/dynamic/himalaya.html

[42] See, e.g., https://www.jetdrift.com/jet-ski-horsepower/

[43] http://www.native-languages.org/travois.htm

[44] https://commons.wikimedia.org/wiki/File:Slaves_Zadib_Yemen_13th_century_BNF_Paris.jpg

[45] Wikipedia

[46] Wikipedia

[47] Qfl547, Wikipedia

[48] Wikimedia

[49] https://commons.wikimedia.org/wiki/File:Ahu-Tongariki-2013.jpg

[50] https://commons.wikimedia.org/wiki/File:Sumerians.jpg, https://www.ancient-origins.net/ancient-places-asia/sumerian-civilization-0011932

[51] Saqib Qayyum, Wikipedia

[52] https://en.wikipedia.org/wiki/John_Bull_(locomotive)

[53] E. A. Wingley, https://royalsocietypublishing.org/doi/10.1098/rsta.2011.0568

[54] http://peakoilbarrel.com/world-energy-2017-2050-annual-report/, https://seekingalpha.com/article/4083393-world-energy-2017minus-2050-annual-report

[55] Photo by Torsten Dedrich on Unsplash, see also: https://e360.yale.edu/features/polar-warning-even-antarctica-coldest-region-is-starting-to-melt

[56] Photo: Josef Friedhuber, https://earthjustice.org/blog/2019-january/arctic-refuge-oil-surveys-put-polar-bears-in-the-crosshairs

[57] Wikipedia

[58] Peltoms, Wikipedia

[59] Courtesy: NASA

[60] https://www.weforum.org/agenda/2019/01/how-an-explosion-of-jellyfish-is-wreaking-havoc/, https://upload.wikimedia.org/wikipedia/commons/8/8f/Blue_jellyfish.jpg

[61] https://upload.wikimedia.org/wikipedia/commons/a/af/Orroral_Valley_Fire_viewed_from_Tuggeranong_January_2020.jpg, see also" https://www.vox.com/science-and-health/2020/1/8/21055228/australia-fires-map-animals-koalas-wildlife-smoke-donate

[62] January 13, 2019, NASA

[63] http://www.millenniumpost.in/kolkata/meeting-to-assist-farmers-deal-with-rain-deficiency-365978

[64] https://static.secure.website/wscfus/8154141/uploads/fbaf02697ab94d23973daf6284b6e131.png

[65] Walter Sigmund, Wikipedia

[66] https://www.astronomynotes.com/solarsys/s4c.htm

[67] https://climate.nasa.gov/news/2616/core-questions-an-introduction-to-ice-cores/, https://cosmosmagazine.com/geoscience/ancient-air-trapped-in-ice/

[68] Leland McInnes; https://commons.wikimedia.org/wiki/File:Co2-temperature-records.svg, https://www.weforum.org/agenda/2018/05/earth-just-hit-a-terrifying-milestone-for-the-first-time-in-more-than-800-000-years

[69] https://www.rsc.org/images/Arrhenius1896_tcm18-173546.pdf, https://www.insblogs.com/uncategorized/climate-change-weve-know-for-some-time/8872

[70] https://upload.wikimedia.org/wikipedia/commons/9/90/CO2-Temp.png

[71] IPCC Third Assessment Report 2001; see also https://www.researchgate.net/figure/GLOBAL-WARMING-POTENTIAL-GWP-FOR-SELECTED-GREENHOUSE-GASES-GHG_tbl1_295513432

[72] Data from: https://www.epa.gov/sites/production/files/2016-05/global_emissions_sector_2015.png

[73] Data from: https://www.epa.gov/ghgemissions/global-greenhouse-gas-emissions-data

[74] https://www.giss.nasa.gov/research/features/200409_methane/core1.gif

[75] https://www.darrinqualman.com/methane/

[76] https://www.sciencedirect.com/science/article/abs/pii/S0301421511001042

[77] https://www.ourenergypolicy.org/growing-poor-slowly-why-we-must-have-renewable-energy/

[78] See also, Write and Conca, 2007; https://blogs images.forbes.com/jamesconca/files/2013/02/HDI-new.jpg

[79] DOE BES 2005; https://www.sandia.gov/~jytsao/Solar%20FAQs.pdf

[80] See e.g., http://geologylearn.blogspot.com/2015/07/formation-of-coal-oil-and-gas.html

[81] https://www.americancraftbeer.com/craft-beers-canary-coal-mine-brewery-closings/

[82] Wikipedia

[83] See e.g., http://geologylearn.blogspot.com/2015/07/formation-of-coal-oil-and-gas.html

[84] https://commons.wikimedia.org/wiki/File:Oiled_Bird_-_Black_Sea_Oil_Spill_111207.jpg, https://bluelife.net.in/2018/03/16/oceans-caught-between-devil-and-the-deep-blue-sea-pun-intended-xi/#jp-carousel-289

[85] https://esajournals.onlinelibrary.wiley.com/doi/full/10.1890/090132

[86] https://en.wikipedia.org/wiki/Oil_shale_in_Estonia#Environmental_impact, https://www.climatechangenews.com/2016/04/07/us-shale-gas-investors-eye-billion-dollar-gamble/, https://insider.si.edu/2014/08/biological-fallout-shale-gas-production-still-largely-unknown/

[87] See e.g., https://friendsoftheearth.uk/climate-change/fracking-facts

[88] https://greencleanguide.com/unconventional-shale-gas-should-india-pursue-it/

[89] See e.g., https://www.usgs.gov/special-topic/water-science-school/science/a-coal-fired-thermoelectric-power-plant?qt-science_center_objects=0#qt-science_center_objects

[90] See e.g., http://needtoknow.nas.edu/energy/energy-sources/fossil-fuels/natural-gas/

[91] See e.g., https://healingearth.ijep.net/energy/history-energy-use

[92] https://www.netl.doe.gov/research/Coal/energy-systems/gasification/gasifipedia/intro-to-gasification

[93] See e.g., ationalgeographic.org/encyclopedia/biomass-energy/#:~:text=Biomass%20is%20organic%2C%20meaning%20it,a%20non-renewable%20energy%20sou

[94] See also https://www.solarschools.net/knowledge-bank/renewable-energy/biomass

[95] https://ourfiniteworld.com/2019/11/14/do-the-worlds-energy-policies-make-sense/, https://oilprice.com/Alternative-Energy/Renewable-Energy/The-Big-Lie-Behind-Global-Energy-Policy.html

[96] https://www.britannica.com/technology/biofuel, https://madison.com/blue-sky-science-could-the-biofuel-cycle-be-made-more/video_12a84a73-4c7b-51f6-9a71-383804887f31.html

[97] Illustration by Sarada Natarajan, see also https://www.researchgate.net/figure/Flex-crops-ethanol-wastes-food-Outside-the-Beltway-2011-Biofuel-vs-Food-Outside-the_fig1_285357699

[98] http://carlosstjames.com/renewable-energy/biodiesel-production-becoming-a-zero-sum-game/

[99] https://www.volker-quaschning.de/datserv/CO2-spez/index_e.php

[100] https://www.holland.com/global/tourism/destinations/provinces/south-holland/the-windmills-of-kinderdijk.htm

[101] Wikipedia, See also https://c03.apogee.net/mvc/home/hes/land/el?utilityname=screc&spc=kids&id=16214

[102] https://commons.wikimedia.org/wiki/File:Paddies_and_wind_turbines_in_India.jpg

[103] Data from GWEC, Wikipedia

[104] https://commons.wikimedia.org/wiki/File:Jigokudani_hotspring_in_Nagano_Japan_001.jpg

[105] https://upload.wikimedia.org/wikipedia/commons/1/17/Image-American_bison_rests_at_hot_spring_in_yellowstone_national_park_1.jpg

[106] See also, https://c03.apogee.net/mvc/home/hes/land/el?utilityname=gru&spc=kids&id=16216

[107] https://commons.wikimedia.org/wiki/File:Pacific_Ring_of_Fire.svg

[108] Data from Think Geo-Energy Research, 2019

[109] http://geosyndicate.com/ge/indianpotential.html

[110] See e.g., https://scijinks.gov/tides/https://theconversation.com/curious-kids-how-does-the-moon-being-so-far-away-affect-the-tides-on-earth-105371

[111] See e.g., https://www.real-world-physics-problems.com/ tidal-energy-for-kids.html

[112] https://openei.org/wiki/Wave_Energy, Ocean Power Technology 2017

[113] Courtesy: NASA

[114] https://science.nasa.gov/earth-science/oceanography/ocean-earth-system/ocean-water-cycle

[115] Tennessee Valley Authority, Wikipedia

[116] https://commons.wikimedia.org/wiki/File:Farniente2.jpg

[117] https://narmadatentcity.info/sardar-sarovar-dam-tourism/

[118] http://sardarsarovardam.org/greenpowerinformation.aspx

[119] Courtesy: NASA

[120] See e.g., https://tasks.illustrativemathematics.org/content-standards/HSG/MG/A/1/tasks/1140

[121] Robert A. Rhode, Wikipedia

[122] Wikipedia, http://www.ez2c.de/ml/solar_land_area/

[123] https://upload.wikimedia.org/wikipedia/commons/c/c8/Olympic_Torch_2010.jpg, https://www.dw.com/en/countdown-to-rio-begins-with-lighting-of-olympic-flame/a-19203169

[124] Painting by Giulio Parigi, https://commons.wikimedia.org/wiki/File:Archimedes-Mirror_by_Giulio_Parigi.jpg

[125] See e.g., https://www.e-education.psu.edu/eme811/node/685

[126] https://solarcooking.fandom.com/wiki/Auroville_Solar_Kitchen?file=Auroville_Solar_Bowl_2.jpg

[127] https://www.energy.gov/eere/solar/power-tower-system-concentrating-solar-thermal-power-basics

[128] Craig Dietrich, Wikipedia, https://www.flickr.com/photos/57385426@N02/9972709085

[129] https://en.wikipedia.org/wiki/Odeillo_solar_furnace#/media/File:Four_en.svg

[130] https://en.wikipedia.org/wiki/Odeillo_solar_furnace#/media/File:Four_solaire_001.jpg

[131] See e.g., https://www.alternative-energy-tutorials.com/solar-hot-water/parabolic-trough-reflector.html

[132] Wikipedia, http://www.ca.blm.gov/cdd/alternative_energy.html

[133] https://commons.wikimedia.org/wiki/File:Abengoa_Solar_(7336087392).jpg

[134] See e.g., https://www.ee.co.za/article/linear-fresnel-systems-future-csp.html

[135] https://worldofrenewables.com/overview-can-the-use-of-thermal-storage-systems-tes-improve-the-efficiency-and-economics-of-concentrated-solar-power-csp/, http://helioscsp.com/reliance-commissions-100-mw-concentrated-solar-power-plant-in-rajasthan-india/

[136] https://commons.wikimedia.org/wiki/File:SolarStirlingEngine.jpg

[137] Zephyris, Wikipedia

[138] YK Times, Wikipedia

[139] See e.g., https://www.ohio.edu/mechanical/stirling/engines/gamma.html

[140] See https://www.ossila.com/pages/solar-cells-theory for details.

[141] See, Michel Bakni, Wikipedia

[142] See e.g., http://hyperphysics.phy-astr.gsu.edu/hbase/Solids/dope.html

[143] Credit: ChemMatters, American Chemical Society, https://www.acs.org/content/acs/en/education/resources/highschool/chemmatters/past-issues/archive-2013-2014/how-a-solar-cell-works.html

[144] https://slideplayer.com/slide/5826416/; Kwang Hoi

[145] Design and In-Depth Optical Studies of Advanced Material Concepts for Luminescent Down-Shifting Coatings for Photovoltaics, Anastasiia Solodovnyk, Ph. D. Thesis, University of Erlangen- Nurnberg (2016)

[146] https://greenecon.net/what%E2%80%99s-pushing-solar-energy-efficiency/energy_economics.html, DOE, Lewis Group, Caltech

[147] https://en.wikipedia.org/wiki/Battery_storage_power_station

[148] https://www.usgs.gov/media/images/water-can-be-reused-produce-hydroelectric-power

[149] Data from, http://solarcellcentral.com/cost_page.html

[150] Rafssbind, Wikipedia

[151] Data from, https://energy.economictimes.indiatimes.com/news/renewable/infographics-global-solar-pv-installed-capacities-by-2030/72992505

[152] https://www.karnataka.com/industry/pavagada-solar-park/

[153] https://www.iihr.res.in/solar-power-integrated-outdoor-mushroom-growing-unit

[154] http://www.inaplanetofourown.net/assets/papers/Beena%20Patel%20-%20Cumulus%20Mumbai%202015.pdf

[155] Illustration by Sarada Natarajan. See e.g., https://news.sky.com/video/inside-the-congo-mines-that-exploit-children-10784310

[156] Quadrennial Technology Review, Department of Energy, USA. September 2015

[157] Data from https://www.fennonen.fi/en/article-page/renewable-energy-problematic-buzzword

[158] https://www.mpie.de/4185352/permanent-magnets-research-for-sustainability, https://www.energy.gov/sites/prod/files/DOE_CMS2011_FINAL_Full.pdf

[159] https://pubs.usgs.gov/of/2004/1050/uranium.htm

[160] Wikipedia

[161] See e.g., https://cen.acs.org/articles/93/i27/Trying-Unleash-Power-Thorium.html

[162] See also: https://en.wikipedia.org/wiki/Nuclear_fission#/media/File:DBP_1979_1020_Otto_Hahn_Kernspaltung.jpg, https://www.britannica.com/science/physics-science/Quantum-mechanics for details.

[163] See e.g., https://openstax.org/books/university-physics-volume-3/pages/10-5-fission for details.

[164] See e.g., https://www.britannica.com/technology/nuclear-reactor/Thermal-intermediate-and-fast-reactors for details.

[165] See e.g. https://www.britannica.com/technology/moderator for details.

[166] https://www.world-nuclear.org/information-library/nuclear-fuel-cycle/nuclear-power-reactors/nuclear-power-reactors.aspx

[167] https://www.nuclear-power.net/

[168] US Energy Information Administration, https://www.energy.gov/ne/articles/nuclear-power-most-reliable-energy-source-and-its-not-even-close

[169] https://pris.iaea.org/PRIS/WorldStatistics/NuclearShareofElectricityGeneration.aspx

[170] https://www.world-nuclear.org/information-library/current-and-future-generation/nuclear-power-in-the-world-today.aspx

[171] https://www.world-nuclear.org/information-library/current-and-future-generation/nuclear-power-in-the-world-today.aspx

[172] https://pris.iaea.org/pris/worldstatistics/nuclearshareofelectricitygeneration.aspx

[173] http://www.barc.gov.in/publications/tb/apsara.pdf

[174] https://www.oecd-nea.org/ndd/pubs/2018/7413-uranium-2018.pdf

[175] https://www.world-nuclear.org/information-library/current-and-future-generation/thorium.aspx#:~:text=World%20monazite%20resources%20are%20estimated,countries%20(see%20Table%20below

[176] https://www.ias.ac.in/article/fulltext/sadh/038/05/0775-0794

[177] https://www.world-nuclear.org/information-library/nuclear-fuel-cycle/nuclear-power-reactors/nuclear-power-reactors.aspx

[178] Wikipedia

[179] See e.g., https://www.britannica.com/technology/nuclear-reactor/Uranium-mining-and-processing for details

[180] Department of Atomic Energy, https://twitter.com/DAEIndia/status/1285845442475913216/photo/1

[181] http://www.ne.doe.gov/genIV/documents/gen_iv_roadmap.pdf

[182] Wikipedia

[183] Stefan Kuhn, Wikimedia; https://phys.org/news/2005-12-safe-nuclear-power-green-hydrogen.html, and references therein.

[184] (See: GIF R&D outlook for Generation IV nuclear energy systems 2009; http://large.stanford.edu/courses/2018/ph241/rojas1/docs/gif-21aug09.pdf; https://www.world-nuclear.org/information-library/nuclear-fuel-cycle/nuclear-power-reactors/generation-iv-nuclear-reactors.aspx)

[185] Courtesy: David Malin, UK Schmidt Telescope

[186] Borb, Wikipedia

[187] Data from IAEA, ENDF, http://hyperphysics.phy-astr.gsu.edu/hbase/NucEne/coubar.html

[188] See e.g., http://hyperphysics.phy-astr.gsu.edu/hbase/NucEne/fusion.html for details

[189] https://www.iter.org/

[190] https://lasers.llnl.gov/

[191] https://www.nextbigfuture.com/, https://commons.wikimedia.org/wiki/File:Inertial_confinement_fusion.svg; https://media.nature.com/original/magazine-assets/d41586-019-00261-3/d41586-019-00261-3.pdf

[192] https://cerncourier.com/a/mumbai-engages-ads-for-nuclear-energy/

[193] Data from https://www.statista.com/, 2018

[194] https://www.world-nuclear.org/

[195] http://www.barc.gov.in/pubaware/images/barriers_radioactivity_release.gif

[196] https://www.aerb.gov.in/

[197] Shigeru23, Wikipedia

[198] Based on https://whatisnuclear.com/waste.html

[199] See e.g., https://www.radioactivity.eu.com/site/pages/Spent_Fuel_Composition.htm for details.

[200] https://www.nrc.gov/reading-rm/doc-collections/fact-sheets/radwaste.html

[201] See e.g., https://whatisnuclear.com/recycling.html for details.

[202] Illustration: Sarada Natarajan; see also, https://www.acs.org/content/acs/en/education/whatischemistry/landmarks/radiocarbon-dating.html

[203] https://en.wikipedia.org/wiki/Hyperthyroidism#/media/File:Blausen_0534_Goiter.png

[204] See e.g., https://www.hzdr.de/db/Cms?pOid=11295&pNid=0

[205] Courtesy: Sandip Basu, Radiation Medicine Centre, Bhabha Atomic Research Centre, Tata Memorial Hospital Mumbai

[206] http://www-naweb.iaea.org/na/news-na/na-water-shortage.html

[207] http://www.barc.gov.in/pubaware/agri_social_inpact.html

[208] https://www.windows2universe.org/earth/Life/cell_radiation_damage.html

[209] https://cerncourier.com/a/the-changing-landscape-of-cancer-therapy/

[210] https://www.facebook.com/iaeaorg/photos/a.131323982061/10154112730192062/?type=3&eid=ARBfJiC1G5dZdU-n7IEQmOS2tOB3hDxHkCTO3OfIKJk4Uno12nMnjOXbbLSwOQW115Rh9hTPSb6aIGQW

[211] https://www.world-nuclear.org/information-library/facts-and-figures/heat-values-of-various-fuels.aspx

[212] Courtesy: NASA

[213] See e.g., R. Dervisoglu, Wikipedia for details

[214] Wikipedia

[215] https://pdfs.semanticscholar.org/242a/68a68dac7aad62a92b9de668d1b36249bab0.pdf

[216] https://en.wikipedia.org/wiki/Hydrail#/media/File:InnoTrans_2016_%E2%80%93_Alstom_iLint_with_Fuel_Cell_Batteries_(29782914176).jpg
https://phys.org/news/2017-11-hydrogen-powered-german-rails.html

[217] https://en.wikipedia.org/wiki/Hydrogen-powered_aircraft#/media/File:HY4_2016-09-29_ueber_Flughafen_Stuttgart.jpg

[218] See also, https://www.energy.gov/eere/fuelcells/hydrogen-production-electrolysis

[219] Credit: Leon Kuhner, KIT,
https://www.kit.edu/img/pi/2019_141_Hydrogen%20from%20Natural%20Gas%20without%20CO2%20Emissions_englisch_72dpi.jpg

[220] Illustration by Sarada Natarajan, https://www.geopura.com/blog/why-we-should-start-using-green-hydrogen-in-2019/

[221] M. Z. Jacobson et el., Joule (2017),
https://www.cell.com/joule/pdfExtended/S2542-4351(17)30012-0

[222] Russell D. Hoffman, http://www.animatedsoftware.com/geni/rh2000ge.htm

[223] https://www.linkedin.com/company/global-energy-network-institute---geni/?originalSubdomain=ca

[224] https://commons.wikimedia.org/wiki/File:Sadi_in_a_Rose_garden.jpg

Index

Accelerator Driven Subcritical System, 253, 254, 255, 256
Agriculture, 14, 51, 54, 56, 57, 58, 62, 64, 66, 83, 95, 130, 198, 277, 310
Agrivoltaics, 196, 198
Animals, 1, 3, 8, 14, 15, 16, 17, 22, 26, 29, 41, 42, 44, 46, 51, 52, 53, 57, 62, 63, 68, 79, 83, 89, 93, 103, 109, 113, 117, 125, 126, 127, 129, 141, 148, 150, 155, 278, 303, 310
Antarctica, 47, 49, 50, 73, 74, 76, 88, 89, 93, 313
Aquavoltaics, 155, 196
Arctic, 73, 74, 75

Bhabha Atomic Research Centre, 226, 281, 283, 284
Bhabha, Homi Jehangir, 224, 225, 226, 227, 228, 253
Biodiesel, 130, 132, 133
Biodiversity, 13, 14, 26, 27, 30, 132, 267, 311
Biofuel, 125, 126, 130, 131, 132, 134
Biomass, 73, 95, 113, 123, 124, 125, 126, 127, 128, 129, 161, 200, 201, 291, 303
Bitumen, 114, 120, 121

Carbon Dioxide, 40, 43, 44, 45, 70, 71, 73, 80, 81, 82, 88, 89, 90, 91, 93, 96, 97, 105, 106, 107, 109, 116, 117, 118, 124, 125, 126, 128, 134, 135, 143, 154, 190, 224, 278, 291, 299, 300, 304, 308, 310, 311
Carbon Monoxide, 110, 111, 118, 124, 157, 291, 299
Carrying Capacity, 54, 60
Chernobyl, 224, 225, 264, 265, 266, 267, 271
Civilization, 10, 32, 50, 51, 54, 56, 57, 60, 61, 62, 63, 64, 65, 66, 69, 99, 161, 303, 308, 311
Climate, 14, 26, 27, 30, 60, 61, 64, 66, 71, 76, 80, 84, 89, 92, 95, 96, 121, 293, 304, 305, 308, 310, 311
Coal, 4, 14, 42, 67, 68, 69, 103, 109, 110, 111, 112, 113, 115, 117, 118, 121, 122, 123, 124, 125, 134, 135, 142, 157, 179, 180, 200, 201, 208, 220, 289, 290, 291, 298, 299, 300
Coal Gasification, 124, 125, 291, 299

Dams, 20, 154, 155, 157, 158, 159, 192, 199
Deforestation, 14, 15, 63, 66, 129, 304
Drought, 18, 51, 60, 62, 78, 81, 155, 304

Earth, 1, 3, 9, 10, 12, 13, 14, 30, 31, 32, 33, 38, 39, 40, 41, 42, 43, 44, 45, 46, 50, 51, 54, 60, 70, 71, 73, 77, 82, 83, 85, 86, 87, 88, 89, 90, 91, 93, 109, 117, 137, 140, 143, 144, 146, 147, 151, 152, 153, 161, 162, 163, 164, 171, 182, 192, 203, 206, 243, 290, 302, 303, 306, 310, 311, 312, 313, 314
Earthquake, 13, 59, 66, 70, 156, 260, 268, 269, 270
Ecosystem, 1, 12, 13, 14, 15, 26, 27, 28, 30, 70, 71, 79, 115, 132, 155, 311, 312

Fast Breeder Reactor, 229, 230, 235, 237, 253, 254, 273
Fission, 208, 209, 210, 211, 212, 213, 214, 215, 216, 217, 218, 228, 229, 240, 241, 253, 254, 259, 267, 271, 272, 274, 275
Flood, 15, 59, 64, 66, 78, 150, 154, 156, 157, 260, 304
Forest, 3, 9, 13, 47, 79, 116, 120, 125, 129, 267, 304, 310, 311
Forest Fire, 13, 79, 304, 310
Fracking, 118, 119, 142, 292

Fuel Cell, 293, 294, 295, 301
Fukushima, 224, 225, 264, 268, 269, 270, 271
Fusion, 39, 162, 213, 243, 244, 245, 246, 247, 248, 250, 251, 289

Geothermal, 109, 140, 142, 143, 144, 145, 146, 147, 201, 303
Glacier, 73, 75, 76, 82, 85, 153, 304
Global Warming, 71, 73, 78, 81, 82, 83, 84, 85, 91, 92, 93, 94, 96, 97, 107, 121, 224, 225, 237, 290, 308, 310
Greenhouse Gas(es), 40, 43, 44, 70, 71, 76, 85, 91, 92, 93, 94, 97, 118, 121, 124, 125, 138, 139, 143, 157, 293
Greenland, 75, 76, 89, 228, 304

Habitat, 13, 18, 73, 79, 80, 82, 113, 132, 304
Heavy Water, 215, 218, 219, 226, 229, 232, 233, 234, 235, 245, 246, 274, 276
Human Development Index, 84, 100, 102, 103, 303, 312
Hydroelectric, 69, 152, 153, 154, 155, 157, 158, 159, 192, 196, 199, 202, 298, 303
Hydrogen, 36, 39, 40, 45, 82, 88, 124, 139, 140, 143, 162, 193, 208, 213, 215, 232, 245, 248, 267, 270, 289, 290, 291, 292, 293, 294, 295, 296, 297, 298, 299, 300, 301, 302, 303, 305
Hydrolysis, 193, 295, 298

Ice Age, 13, 70, 85, 86
Industrial Revolution, 25, 56, 66, 67, 68, 69, 71, 77, 95, 103, 109, 311
IPCC, 95, 97, 304

Mahatma Gandhi, 309, 310, 311
Methane, 28, 40, 41, 76, 93, 94, 95, 96, 111, 118, 154, 290, 299, 300, 304
Molten Salt Reactor, 237, 238, 239, 242, 254
Moon, 9, 31, 32, 39, 40, 46, 47, 147

Natural Gas, 68, 106, 109, 110, 113, 114, 115, 116, 117, 118, 121, 122, 124, 134, 289, 290, 291, 292, 298, 300, 303, 304
Nitrous Oxide, 93, 94, 95, 96, 118, 135
Nuclear Energy, 71, 109, 220, 221, 222, 223, 225, 227, 228, 237, 242, 257, 264, 303, 304
Nuclear Power, 109, 139, 191, 200, 202, 205, 217, 219, 220, 223, 224, 225, 228, 233, 236, 261, 264, 268, 271, 292, 298, 299, 306, 307, 310
Nuclear Radiation, 277, 280
Nuclear Reactor, 69, 212, 214, 216, 217, 219, 224, 226, 235, 236, 240, 242, 253, 255, 257, 259, 260, 264, 265, 268, 271, 275, 299
Nuclear Safety, 257, 264, 271, 306
Nuclear Waste, 225, 238, 254, 256, 257, 271, 273, 274, 275, 276

Ocean, 7, 10, 28, 29, 30, 31, 40, 41, 43, 46, 50, 62, 75, 78, 82, 83, 88, 89, 91, 93, 95, 107, 113, 147, 151, 152, 153, 231, 245, 304, 308, 313
Oil, 15, 42, 67, 68, 69, 79, 103, 106, 109, 110, 113, 114, 115, 117, 118, 119, 120, 121, 122, 124, 125, 132, 133, 134, 283, 289, 290, 291, 292, 293, 298, 303, 304
Oil Sand, 120, 121
Ozone, 44, 94, 163

Palm Oil, 15, 132, 133, 310
Pebble Bed Reactor, 239, 241
Plastic, 29, 30, 113, 166, 175, 184, 190, 202, 292, 299, 312
Plutonium, 213, 217, 218, 235, 253, 254, 271, 272, 273, 275
Pollution, 14, 26, 27, 29, 30, 118, 138, 164, 281, 290, 293, 306, 308, 311
Pressurised Heavy Water Reactor, 229, 236, 237, 246
Pressurised Light Water Reactor, 216
Pumped-up Hydroelectric Storage, 154, 157, 159, 192, 193

Shale (oil/gas), 118, 119, 120
Silicon, 181, 182, 183, 184, 185, 188, 189, 194, 202, 240, 299

Slave(ry), 54, 55, 56, 57, 58, 63
Solar Cell, 180, 182, 184, 185, 186, 187, 188, 189, 190, 191, 193, 198, 203
Solar Energy, 128, 149, 161, 162, 164, 167, 172, 173, 180, 185, 201, 303
Solar Farm, 142, 155, 191
Solar Panel, 155, 159, 190, 191, 192, 196, 197, 198, 199, 202, 203, 306
Solar Photovoltaic, 164, 180, 190, 193, 195, 202, 298, 301, 304
Solar Power, 109, 140, 154, 155, 156, 159, 169, 170, 176, 179, 180, 185, 190, 191, 193, 194, 195, 196, 198
Solar Thermal, 164, 165, 167, 171, 174, 177, 179, 180, 193, 304
Sulphur Dioxide, 116, 124, 135, 143, 157, 283, 291, 300
Sun, 2, 3, 10, 11, 31, 32, 39, 40, 44, 46, 51, 59, 87, 88, 109, 126, 137, 147, 161, 162, 163, 164, 166, 167, 168, 170, 171, 179, 185, 191, 213, 243, 244, 245, 248, 290, 306, 313
Thorium, 45, 109, 140, 205, 207, 210, 219, 227, 228, 230, 235, 238, 240, 241, 247, 253, 254, 256, 271, 273, 306
Three Mile Island, 225, 264, 265, 271
Tidal, 109, 147, 148, 149, 150, 292, 298, 303
Tokamak, 248, 250
Tsunami, 13, 268, 269, 270

Uranium, 36, 45, 109, 135, 140, 205, 207, 208, 210, 213, 217, 218, 219, 226, 227, 228, 229, 230, 231, 232, 233, 234, 235, 238, 240, 241, 247, 248, 253, 254, 266, 271, 273, 274, 275, 276, 306

Wave, 2, 151, 161, 298
Wind(mill), 2, 40, 50, 71, 109, 118, 137, 138, 139, 140, 142, 148, 154, 159, 161, 171, 175, 191, 192, 199, 202, 292, 295, 298, 301, 303, 305, 306

www.ingramcontent.com/pod-product-compliance
Lightning Source LLC
LaVergne TN
LVHW082021150826
845671LV00006B/227
* 9 7 8 9 8 1 1 2 3 3 4 7 0 *